# WHY OXYTOCIN MATTERS

*About the author*

Kerstin Uvnäs Moberg, MD, PhD, is recognised as a world authority on oxytocin. Her research has taken place at the famed Karolinska Institute in Stockholm, and at the Swedish University of Agricultural Sciences in Uppsala, where she is Professor of Physiology.

She is the author of more than 450 scientific papers and the books *She and He*, *The Oxytocin Factor*, *The Hormone of Closeness* and *Oxytocin: the Biological Guide to Motherhood*. She co-authored *Attachment to Pets*.

Dr Uvnäs Moberg lectures widely in Europe and the United States. Her work has been influential in a variety of fields, including physiology, women's health, obstetrics, psychology, animal husbandry, physical therapy, paediatrics and child development.

She is the mother of four children, and she lives in Djursholm, Sweden.

These books have been provided by the Strong Start Team working in partnership with Milk and You.
Strong Start are qualified Early Years professionals commissioned by Public Health to provide the universal part of the Children's Centre Services offer for West and North Northants
strongstartteam@westnorthants.gov.uk
07880136070
Milk&You are Public Health commissioned breast feeding peer support volunteers trained to offer guidance on infant feeding.
07949353423

*Why Oxytocin Matters (Pinter & Martin Why It Matters 16)*

First published by Pinter & Martin Ltd 2019



ISBN 978-1-78066-605-1

Also available as an ebook

Pinter & Martin Why It Matters ISSN 2056-8657
Series editor: Susan Last
Index: Helen Bilton

Illustration page 19 by Hannah Wagner

British Library Cataloguing-in-Publication Data

A catalogue record for this book is available from the British Library.



Set in Minion

Printed and bound in the EU by Hussar

This book has been printed on paper that is sourced and harvested from sustainable forests and is FSC accredited.

Pinter & Martin Ltd
6 Effra Parade
London SW2 1PS

pinterandmartin.com

# Contents

# Introduction

This book is for parents, and all those who help and support mothers and fathers when a baby is born. Its purpose is to present information about oxytocin: a substance that plays a very important biological role in parents and their newborn babies.

I belong to a generation of female scientists, whose scientific careers were completely changed after becoming a mother. I was a medical doctor who became a researcher in the field of physiology/pharmacology; that is, the understanding of how the different functions of the body are organised and controlled, and how drugs and other substances influence these processes. As a consequence of my experiences of giving birth to and breastfeeding four children, I became convinced that innate biological systems that help and support mothers and fathers in their role as parents must exist.

The longer I have worked to investigate these systems, the more I have come to admire the smartness of the biological adaptations that I discuss in this book. These evolutionary processes not only helped parents when the human species was young, but also continue to do so now, if we can acknowledge them and allow them to be expressed. Knowledge of these inborn competencies, which automatically adapt parents' behaviour and even optimise their bodily functions, can make parenthood easier and less stressful for new parents.

In the course of my work, it soon became apparent that newborn babies can be influenced early in life: their future behaviour and bodily functions can be adapted to the environment they are born into. Tough environments make babies tougher, while friendly and supportive environments make babies friendlier and calmer, and perhaps happier and healthier as adults.

I discovered that oxytocin, a hormone known to be involved in birth and breastfeeding, has a far wider and more important role. Rather than just stimulating the mechanical aspects of birth and breastfeeding, it optimises the mother's behaviour and physiology in these situations. So oxytocin – or the oxytocin system, as I prefer to call it – both promotes reproduction in a broad sense, and, more specifically, helps mothers to create and feed their babies, as well as helping both mothers and fathers to care for the newborn and allow babies to grow and develop in an optimal way. The oxytocin system is the 'director' of the whole story that I am going to tell. It acts in the brain, it acts in the blood circulation and it even acts in individual cells to promote most aspects of parenthood.

## The mother hormone

Oxytocin was first 'detected' at the beginning of the 20th century when it was shown to cause uterine contractions during birth and the ejection of milk during breastfeeding. As oxytocin was originally considered to be a hormone that helps mothers during labour, birth and breastfeeding, it was labelled 'the mother hormone'. Unfortunately this label meant that further research on oxytocin was not considered very interesting or important, and consequently research into the oxytocin system was delayed for a long time. It was not until many years later that oxytocin was found to induce important parallel behavioural and physiological effects that help mothers adapt to motherhood.

## The hormone of love

Towards the end of the 20th century there was a resurgence of interest in oxytocin, and it soon became an extremely popular substance for scientists to study. It became clear that oxytocin not only circulated in the bloodstream, but was also present in nerves in the brain and could influence brain function. We learned that oxytocin was as much a male as a female hormone, and that it was present in individuals of all ages. It emerged that oxytocin not only stimulates interaction and bonding between mothers and infants, but also social interaction and bonding between individuals in general, for example between friends or partners, and this bonding might be linked to general wellbeing. Journalists reported the findings, producing masses of articles about oxytocin and implying that widespread distribution of this miraculous substance would make us all friendly, happy and attractive. Oxytocin became known as 'the hormone of love'.

## The hormone of health and life

However, researchers have also demonstrated that oxytocin can exert many other extremely important physiological effects by actions both in the brain and elsewhere, including reducing the sensation of pain, reducing levels of fear and stress and counteracting inflammation. In addition, oxytocin has been shown to stimulate growth and restorative processes. It may even stimulate the division of healthy cells and stop unhealthy cells from growing. Taking all these effects into consideration, it's obvious that oxytocin plays a fundamental and important role in all of us, because the sum of all these effects is that oxytocin promotes health and in a broader sense life. Therefore, oxytocin should be labelled 'the hormone of health and life'. This is why oxytocin matters.

In this book we will begin with some basic information about oxytocin before looking at the more specific effects of oxytocin during the different stages of parenthood. We will look at how oxytocin is released in mothers during pregnancy, labour, birth, skin-to-skin contact after birth and breastfeeding, and how it influences both mothers' and fathers' interactions with their infant later on, and how it affects the bonding that takes place between them.

Large amounts of oxytocin are released in the mother in connection with birth, during skin-to-skin contact after birth and during breastfeeding. Oxytocin is also released in both mothers and fathers when they are in close contact with the baby. What is now becoming clear is that unborn and newborn babies also have their own oxytocin system, which releases oxytocin during these events. Consequently, oxytocin induces a myriad of effects in the whole family in the period around birth.

Not only is the oxytocin system acutely activated during the period around birth, but it is also boosted more than in any other situation in life. In fact, the oxytocin-linked stimulation of positive social interactive behaviour and decrease in stress levels that occurs in parents and infants at this time will, to a certain extent, become sustained or imprinted. In this way the oxytocin-linked effects that are induced during birth and breastfeeding may be of importance for wellbeing and health far into the future.

As groundbreaking and important as this knowledge is – that the positive effects of oxytocin on health and future relationships may be sustained in parents and their children in the longer term – there are obvious challenges to consider. What happens to the oxytocin system and its effects in a world where spontaneous, vaginal labour is becoming less common, and breastfeeding is declining? Will the reduction in oxytocin exposure lead to diminished wellbeing and health in mothers, fathers and their babies, and perhaps future generations of humans? Or are there ways to substitute for this loss? These questions need to be addressed and discussed.

## Summary

- Oxytocin was originally known for its role in labour and breastfeeding and was thought of as 'the mother hormone'.
- Because oxytocin was found to stimulate wellbeing and the formation of various kinds of relationships between individuals, it was thereafter labelled the 'hormone of love'.
- As oxytocin also decreases stress levels and promotes processes related to growth and healing, it should really be called 'the hormone of health and life'.
- Large amounts of oxytocin are released in connection with birth and breastfeeding in both parents and babies. The strong oxytocin effects induced during this period may become sustained and thereby influence wellbeing and health in the long term.

# 1 Oxytocin

Most women have heard about oxytocin. Some men have also heard about it, particularly nowadays, since giving birth to babies and taking care of children is much more of a shared task than it used to be. So you may know that oxytocin is a hormone that contracts the uterus during labour, and that it helps mothers' milk ejection during breastfeeding. You may also know that many women are given extra oxytocin as an intravenous drip during labour; either to induce (start) or augment (boost) contractions. Some of you may know that oxytocin is often given after labour to help the uterus contract, and reduce blood loss after birth. Some parents may have heard of oxytocin as 'the hormone of love', and that oxytocin in a miraculous way may make you happy, attractive and successful. Even if these sorts of statements are largely exaggerated, oxytocin does have many positive effects that help mothers and fathers become parents, but which may also influence the wellbeing and health of both parents and baby in the longer term.

## What is oxytocin?

The British physiologist Sir Henry Dale discovered oxytocin at the beginning of the 20th century, more than 100 years ago. He found that the pituitary gland (one of the most important hormone-producing glands in the body, which is located just underneath the hypothalamus, an ancient part of the brain), contained a substance that caused uterine contractions and milk ejection when it was injected into cats. In fact, oxytocin was one of the very first human hormones to be discovered. (A hormone is a substance that is released into the circulation and then reaches another organ, where it binds to receptors and exerts effects).

### *The chemical structure of oxytocin*

Soon the search for the structure of oxytocin was initiated. A new and important phase in the history of oxytocin began in the 1950s when an American researcher, Vincent du Vigneaud, found that oxytocin is a small protein or polypeptide consisting of nine amino acids (amino acids are the building blocks of all proteins).

As soon as the identity of the nine amino acids of the oxytocin molecule was established, and the order in which they were linked to each other became clear, it was possible to develop techniques by which oxytocin could be produced or synthesised. Once these processes were perfected, large amounts of oxytocin could be produced and it could be used clinically. It may seem unbelievable, but synthetic oxytocin produced using these chemical techniques has been available and been given as an intravenous drip to women in order to stimulate uterine contractions during labour or prevent bleeding after birth, and as a nasal spray to facilitate milk ejection during breastfeeding, since 1960 – almost 60 years!

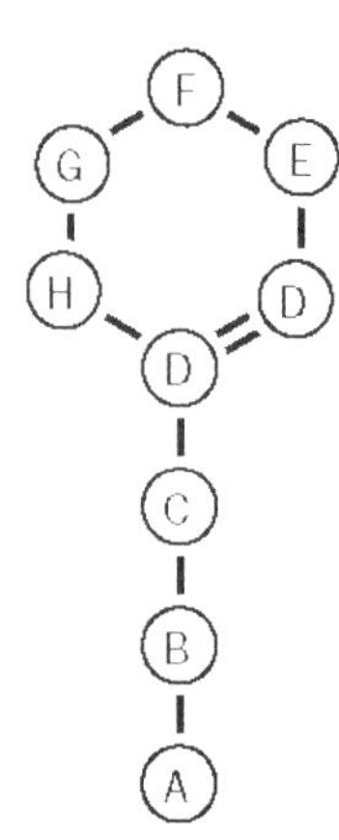


*Fig 1.* Simple illustration of the chemical structure of oxytocin.

*Natural (endogenous) oxytocin and synthetic (exogenous) oxytocin*

Many people, including those who work in maternity wards and give infusions of synthetic oxytocin to mothers during or after labour, believe that synthetic oxytocin (also sometimes called Pitocin or Syntocinon) is different from the natural or endogenous oxytocin produced in the body. But since the synthetic (exogenous) oxytocin is made as a copy of the natural oxytocin which is produced by the body (endogenous oxytocin), synthetic oxytocin is, from a chemical point of view, completely identical to the oxytocin produced within the body.

American researcher Vincent de Vigneaud first identified the structure of the oxytocin molecule and also that of a related hormone, vasopressin. This was considered to be such an important scientific breakthrough that he was awarded the Nobel Prize for Chemistry in 1955. Sir Henry Dale, who discovered oxytocin, had received a shared Nobel Prize in Medicine and Physiology in 1936 for his groundbreaking work in chemical transmission.

What Sir Henry Dale and Vincent du Vigneaud didn't know, when they carried out their original work, was that within 50 years there would be renewed and expanding interest in oxytocin. This later work led to many more functions of oxytocin being identified in addition to the original effects on labour and milk ejection, and it was found that oxytocin is not just a female hormone, but also causes important effects in both men and women, of all ages.

## Where is oxytocin produced?

Oxytocin is produced in two small cell groups called the supraoptic nucleus (SON) and the paraventricular nucleus (PVN), which are located in a very basic part of the brain, the hypothalamus. Oxytocin is produced in nerve cells in these groups and is moved (along long extensions called axons) from the hypothalamus to the posterior part of the pituitary gland, from where it is released into the bloodstream (circulation). It then reaches the uterus or the mammary glands (breasts) via the bloodstream to exert its classical physiological actions: stimulating uterine contractions during labour, and stimulating milk ejection during breastfeeding. When oxytocin is released during labour and breastfeeding, it is released in a very specific way into the bloodstream: in short pulses that in time correspond to spikes of electrical activity that occur in the oxytocin-producing cells in the SON and PVN.

### *Oxytocin acts in many different ways*

A chemical substance is called a hormone (in this case oxytocin) if it is released into the bloodstream from one organ (in this case the posterior pituitary) and then transported via the bloodstream to another organ (in this case the uterus or breasts), where it exerts actions.

Oxytocin, however, is not just a hormone. It also acts in

other ways. In the 1980s, researchers discovered that oxytocin-containing nerves emanating from a special group of cells in the PVN, called parvocellular neurons, and extensions from the nerves connecting the PVN and SON with the pituitary, reached many important regulatory areas within the brain. Oxytocin nerves reach areas involved in the control of social interaction, wellbeing/reward, stress and other basic functions. By sending information via nerves in the brain, oxytocin acts as a neurotransmitter or a chemical signalling substance.

Oxytocin may also act in a more direct way within the brain. When the oxytocin-producing cells in the PVN and SON of the hypothalamus are exposed to very strong stimulation, for example during birth and breastfeeding, oxytocin is released in large amounts from the part of the oxytocin-producing nerve that is within the hypothalamus, the cell body and dendrites (short extensions). This means that oxytocin is released directly into the surrounding tissues, so neighbouring areas in the brain may become flooded with oxytocin and consequently oxytocin effects may be induced.

Recent research has shown that oxytocin is also produced in peripheral organs such as the cardiovascular and gastrointestinal system. These peripheral oxytocin systems exert local actions in the tissues in which they are produced and are activated in response to local triggers. However, it is likely that the peripheral oxytocin systems are in some way linked to the central or 'mother' oxytocin system in the hypothalamus, and that all parts of the oxytocin system are somehow connected to each other and that their activities can be synchronised.

## What does oxytocin do?

The results of the first experiments in which Sir Henry Dale showed that oxytocin was involved in the processes of labour and birth suggested that oxytocin was a female reproductive hormone. This, however, is not the case. It is as much a male as a female hormone, and it is present in people of all ages, from babies in the womb to elderly people. It just happens to be of extreme importance and released in very high quantities during birth and breastfeeding.

*Social interaction*

The effects of oxytocin on social interaction are not restricted to the interaction between mothers and babies. Oxytocin also stimulates many other types of friendly social interaction between individuals, within couples, families or groups. It may also help individuals to bond or attach to each other. The effect is twofold. Oxytocin both stimulates friendly social interaction and is released in response to friendly social interaction. In this way a positive oxytocin spiral may be created by good relationships.

Oxytocin has some other very important effects. For example, it decreases levels of anxiety or fear, and it may induce wellbeing. It also lowers stress levels (by reducing blood pressure and suppressing the stress hormone cortisol). It reduces pain and inflammation. The processes of digestion, metabolism, restoration, healing and growth may be optimised by the presence of oxytocin. Obviously all these effects together point in the same direction: they are positive for individuals and will promote health and, in the long term, life.

To simplify the broad spectrum of oxytocin effects, they can be divided into four main categories:

- Stimulation of contraction of muscles in the womb and breasts
- Stimulation of social interaction
- Decrease of inflammation and stress
- Stimulation of growth and healing

## Oxytocin effects on the nervous system

Some of the basic effects of oxytocin that are referred to repeatedly in this book are those linked to the inhibition of stress and the stimulation of processes related to healing and growth. These effects of oxytocin are exerted by influencing the function of the HPA axis and the autonomic nervous system. A very short summary of the structure and function of these systems is below, to make it easier to understand these basic functions of oxytocin.

The 'stress' system consists of two basic parts, the hypothalamo-pituitary-adrenal (HPA) axis and the sympathetic part of the autonomic nervous system.

The 'anti-stress' system consists of the oxytocin system and the parasympathetic nervous system.

### *The HPA axis*

Levels of the stress hormone cortisol are regulated by the HPA axis. Corticotrophin-releasing hormone, or CRF, which is produced in the PVN of the hypothalamus, where oxytocin is also produced, stimulates the release of a hormone from the pituitary, which in turn stimulates the release of cortisol.

### *The nervous systems*

The autonomic nervous system consists of two branches, the sympathetic branch and the parasympathetic branch. For the purpose of this book it is sufficient to say that the sympathetic nervous system controls functions related to activity or stress, whereas the parasympathetic nervous system regulates functions linked to digestion, storing of energy and growth. Oxytocin decreases activity in the sympathetic nervous system and increases activity in the parasympathetic (vagal) nerve.

Both the 'stress' system and the 'anti-stress' system are complex and are composed of many different mechanisms and effects which occur at many different levels in the brain and the body. For the sake of simplicity I refer to them here as the 'stress' system, or CRF, and the 'anti-stress' or oxytocin system.

There is a very important relationship between the 'stress' system and the 'anti-stress' system. Oxytocin inhibits the stress axis or the secretion of CRF, and the stress axis inhibits oxytocin secretion/effects. In other words, when stress levels are high, oxytocin release and the knock-on effects decrease. Conversely, if activity in the oxytocin/'anti-stress' system is high, oxytocin decreases the secretion and effects of CRF. When the 'stress' system is in charge, the biological pattern of fight and flight prevails; but when the 'anti-stress' or oxytocin axis is dominant the calm and connection response prevails.

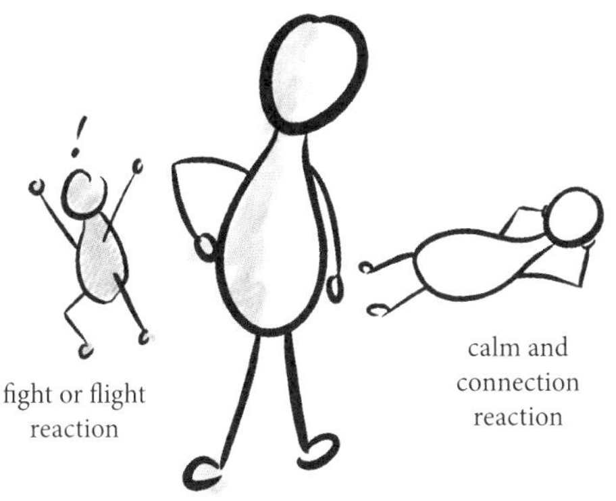


*Fig 2.* The fight and flight reaction and the calm and connection system

This basic antagonism between the 'stress' and 'anti-stress' systems will come up many times in this book. Inhibiting stress responses and increasing activity in the 'anti-stress' system is one of the most important effects of oxytocin. It is one of the main effects exerted by oxytocin in all of the situations we will look at: pregnancy, labour, birth, skin-to-skin contact and breastfeeding.

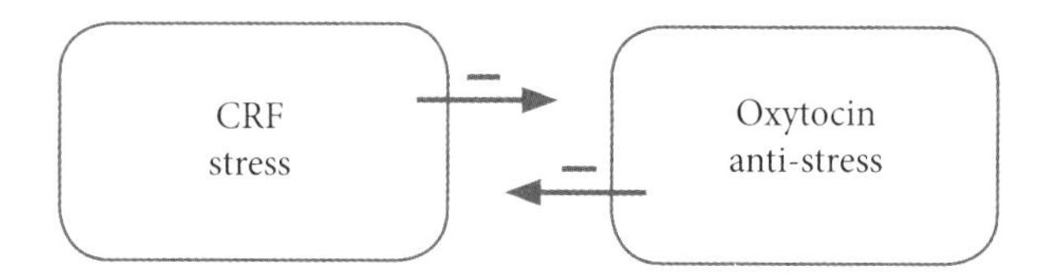


*Fig 3.* The antagonism between stress/anti-stress systems

## Oxytocin coordinates patterns of effects

One important aspect of the oxytocin system is that it coordinates different, separate effects in a meaningful pattern. All mammals use oxytocin to give birth and to give milk to their offspring. But in addition oxytocin stimulates a vast number of psychological, behavioural and physiological effects that are associated with and facilitate all aspects of birth and lactation.

### *Care*

Mothers take care of their newborns and bond with them. These caring behaviours, as well as the bonding between mothers and their young (and vice versa), are stimulated by oxytocin. If oxytocin is administered to mammalian females, a behavioural pattern called 'maternal behaviour' can be triggered. Of course, caretaking behaviours differ between species depending on their physical appearance, how many offspring they give birth to, how mature the offspring are when they are born and the environment in which they live. Polar bear mothers take care of their newborns differently from sows or rats. But the intention is the same in all mammalian mothers: to help their offspring to survive. So oxytocin not only stimulates the physiological processes of giving birth and giving milk, but also helps with the psychological and/or behavioural aspects of motherhood.

*Protection*

One important aspect of maternal behaviour or motherhood is to protect and defend offspring from dangerous and threatening situations and individuals. Mothers can be extremely aggressive in order to protect their offspring and this effect is also stimulated by oxytocin. Paradoxically, oxytocin does not only induce caring behaviours, but can also induce aggression and defensive reactions when needed. It all depends on the situation. We can think of oxytocin as serving life: therefore, it can stimulate caring or protecting behaviours depending on the circumstances. In calm and familiar surroundings caring behaviours are expressed, but if the environment is experienced as unsafe or threatening, aggressive and protective behaviours will be activated.

*Other physiological adaptations*

During birth and lactation a mother's oxytocin adapts her digestive and metabolic processes so that she becomes more 'energy efficient': growing a baby and producing milk requires extra calories. Stress levels are also often decreased during these periods.

## Effects of oxytocin throughout life

As we have seen, oxytocin has a wide spectrum of effects, ranging from stimulation of social interaction and reduction of stress levels to stimulation of processes linked to growth and healing. All these effects occur during the different phases of life, although they may be expressed in slightly different ways or at different levels of intensity. There are, however, some specific actions of oxytocin that may be restricted to certain phases of life. A few examples are given below.

*Oxytocin in the baby*

Oxytocin is produced in the unborn baby and probably plays a role in stimulating processes related to maturation and growth, as well as the development of different aspects of social behaviour.

At birth oxytocin helps the baby by decreasing levels of pain, just as it does in the mother. It also minimizes tissue damage caused by birth, for example in response to low levels of oxygen (hypoxia). In this case the anti-inflammatory actions and restorative functions of oxytocin come into play.

*Oxytocin and fertility*

Oxytocin levels surge at puberty in girls/women due to rising levels of the female sex hormone oestrogen, which increases the release of and effects of oxytocin. In addition, oxytocin levels surge during ovulation during each menstrual cycle. This is probably the reason why women are more social and interactive at this time.

Oxytocin also stimulates fertility in more local and specific ways, for example by causing eggs in the ovaries to mature, stimulating the transport of the eggs through the fallopian tubes and helping the fertilised egg to implant in the wall of the womb.

Men's fertility is also influenced by oxytocin. In fact, the production of sperm is positively influenced by oxytocin.

Oxytocin is also linked to sex and is released in large amounts during orgasm in both women and men. Oxytocin may contribute to feelings of love and happiness during intercourse, and also to the calm and bonding between the partners that may follow afterwards.

*Oxytocin and ageing*

Most processes in the body slow with age, and the oxytocin system is no exception. In women, activity in the oxytocin system and oxytocin levels decrease with age, in part as a consequence of menopause and decreasing oestrogen levels. However, oxytocin levels and function also decrease with age in men, suggesting that other factors are also involved in the slowing down of the activity in the oxytocin system in older people.

What are the consequences of this age-related decrease in the function of the oxytocin system? It is likely that it is an important factor in some of the mental changes that occur with age, such as a decreased interest in social contact and perhaps also increased levels of anxiety and higher stress levels.

A decline in oxytocin activity may also be involved in the physical ageing process. There are many studies showing that oxytocin contributes to decreased inflammation and increased regeneration in tissues such as the heart and also muscles in general. Lowered oxytocin levels may therefore contribute to the development of osteoporosis and the atrophy of skin and mucosal linings that develops with age. Perhaps lowered activity in the oxytocin system contributes to the ageing and atrophy of muscles in the heart and elsewhere. In the future, oxytocin might even be used as an 'anti-ageing' drug; it has been shown to prevent osteoporosis and rejuvenate the lining of the vagina in menopausal women.

## Oxytocin, social interaction and health

One reason for lowered levels of oxytocin and effects of oxytocin in older people might be that they are less engaged in social contact. This may be because they don't want to be social, or simply because they live alone.

It is difficult to measure oxytocin levels, because the different techniques used don't always give comparable results. But some studies show that levels of oxytocin rise when individuals are interacting in a friendly way. From this perspective it is of interest that people who live together have higher levels of oxytocin than those who live alone. Social contact or touch is one of the best ways of increasing oxytocin levels.

People who live together in positive relationships also experience better health and longer life expectancy. Since oxytocin is linked to health, the increase in oxytocin levels/function caused by positive and friendly social interaction may contribute to the better health and longer life expectancy of these people.

However, remember that oxytocin release and its effects are sensitive to the quality of the environment and the type of interaction. If social interaction is linked to threat or fear, oxytocin levels may decrease or the aggressive and defensive pattern of oxytocin-induced behaviour may be activated. In such situations the oxytocin-linked positive effects on health will fail to appear.

Positive relationships do not necessarily have to involve only humans. There are other types of relationships that can increase oxytocin levels and improve health. Being attached to a pet animal, for example, creates the same stimulation of the oxytocin system and thereby promotes health.

## Oxytocin spray

It is not particularly easy to give people additional oxytocin. If it is given as a pill, it is broken down in the stomach and intestines and is not absorbed into the bloodstream in significant amounts.

However, oxytocin can be given as a nasal spray. Although the effects of giving oxytocin in this way have been questioned, there are clearly some interesting positive effects on social interaction, anxiety and stress levels. Positive effects of oxytocin have been noted in individuals with autism, schizophrenia, anxiety and post-traumatic stress disorder (PTSD). I mention this as an interesting aside, but in the rest of the book our focus is on the effects of 'natural' or endogenous oxytocin, released during pregnancy, birth, breastfeeding and close contact between parents and their baby.

## Summary

- Oxytocin is produced in the hypothalamus and is released into the circulation to act as a hormone. It is also released via nerves in the brain to act as a neurotransmitter and influence brain function.
- Oxytocin decreases activity in the HPA axis and sympathetic nervous system, and increases the effect in the parasympathetic/vagal nervous system.
- When activity in the oxytocin system is high, stress levels are low. When activity in the 'stress' system is high, activity in the oxytocin system is low.
- Oxytocin may coordinate several separate behavioural and physiological effects into biologically meaningful patterns.
- Oxytocin is present in females and males and in young and old individuals. It induces certain effects irrespective of sex and age, but also has some age-specific effects. Oxytocin levels/function decrease with age.
- Oxytocin levels are elevated in response to touch, warmth and good relationships.
- High oxytocin levels/functions are linked to health.

# 2 Oxytocin during pregnancy

The creation of a baby is of course a miracle, but in the physical world it is regulated and promoted by physiological processes that take place inside the mother's womb to allow the fertilised egg to develop first into an embryo and later on into a baby. As the baby grows and develops inside the womb, the mother's brain and body are affected by a cocktail of hormones that support the pregnancy and prepare her for taking care of her baby once it is born.

Normally we think of the female sex hormones oestrogen and progesterone as the cause of the physical and mental adaptations that occur in mothers during pregnancy. These female hormones are produced in the ovaries during the first phase of pregnancy, but later on the placenta takes over as the main site of production of oestrogen and progesterone. Levels of these hormones increase enormously during pregnancy and as a consequence a multitude of other functions in the body and brain are triggered to facilitate pregnancy.

## The link between oestrogen and oxytocin

The oxytocin system is also of great importance during pregnancy. Many of the changes that occur during pregnancy are orchestrated by oxytocin, or rather by the oxytocin system, consisting of hormonal effects of oxytocin via the bloodstream, central effects of oxytocin released from nerves in the brain and local effects produced by oxytocin released from peripheral organs/cells. This is possible because of a close interaction between oestrogen and the oxytocin system. Oestrogen, and to a lesser extent progesterone, increases the release of oxytocin into the circulation, as well as into certain areas of the brain. It also increases the function of oxytocin receptors in specific areas in the brain and the body. Levels of oxytocin increase slowly during pregnancy and by the time of birth are three to four times higher than before pregnancy. This rise in oxytocin levels runs in parallel with the rise in oestrogen levels.

## A psychological transition period

Pregnancy is a period of transition for the mother. Not only does she physically change, but she also prepares mentally for the shift from being one individual to two (or possibly more) separate beings. Giving birth to a baby, especially a first baby, means that a woman, from one moment to another, becomes a mother with all the consequences of motherhood. She has to become familiar with the idea of loving another individual more than she might think is possible – perhaps even more than herself – and of putting another individual's needs before her own. No longer will she be responsible only for herself; in the future she will be responsible for another individual day and night, perhaps when she is more tired than at any time in her life due to lack of sleep.

Oxytocin in the brain helps the mother in this process by opening up her emotions and increasing her wish for – and potential for – social interaction. She becomes more socially interactive during pregnancy, which allows her to communicate with and seek and receive information, support and help from her partner, parents, other members of the family and friends. Old memories become accessible and it may seem as if old conflicts can be more easily resolved and new relationships formed. All of these changes are of great importance for the mother-to-be, since she will need help and protection once her baby arrives.

*Transfer of women's knowledge*

Transfer of knowledge and experience from older women (relatives or friends) to mothers-to-be has historically been of extreme importance for pregnant women. In this way, mothers down the generations have been given invaluable information about the process of pregnancy, birth and breastfeeding. Today the structure of society has changed, and often families live far apart from each other, so this kind of transfer of information is not always available to pregnant women. However, the need to talk and seek information is still there. Nowadays young female friends, midwives and information from the internet have taken over the role of the informant. Perhaps most valuable are birth preparation or discussion groups where women with a variety of experiences can share their knowledge and learn from others.

*Trust facilitates learning*

The effect and value of the information women receive may vary depending on the person who is delivering the information. When a person that you trust or love gives you information, the oxytocin system is activated and levels of oxytocin are high. In this situation, the information you receive will be taken in more efficiently, as it is linked to feelings of safety and trust. When information is obtained from an unfamiliar person or, in the case of the internet even a non-living system, the connection between the new and often excellent information and the experience of trust and safety may be lost. In the absence of love, trust and familiarity with the 'messenger', information received may become connected with fear and worry and not fully taken in and integrated.

*The start of attachment*

Oxytocin also helps to shift the mother's interest from the surrounding material world to her future baby and its needs. In fact, the mother starts to think of, talk to and even bond with her baby during her pregnancy. The process of bonding, which I describe in more detail later on in this book, is a biological process by which individuals become tied together in a positive way. The bonding process starts when the baby is in the womb, but develops further in connection with birth. Oxytocin, which is continuously released during pregnancy, birth and breastfeeding, facilitates the bonding process.

## Metabolic changes in order to promote growth

During pregnancy the mother's body is adapted to allow and facilitate growth of the baby. This takes place in several steps and by activation of several mechanisms, many of them linked to increased activity in the oxytocin system.

*Weight gain*

Oxytocin-linked metabolic adaptations start very early during pregnancy. Many mothers notice that they start to gain weight during the early stages of pregnancy. Women who have already had children may notice in subsequent pregnancies that they put on weight right at the start of pregnancy, when the baby is still only the size of a thumb and can't really be contributing to their weight gain! Women may even notice that they gain weight despite eating less than normal if they have lost their appetite or suffer from morning sickness. In the early part of pregnancy the increase in the mother's weight is caused by her own metabolism, as will be discussed below; later on, both mother and baby contribute to pregnancy weight gain, with the baby contributing more during its phase of very rapid growth.

*The first part of pregnancy: focusing on storing energy*

A mother needs to put on weight at the beginning of pregnancy in order to have stored energy to use later on during pregnancy (and breastfeeding), when the daily need for calories increases due to the growth of the baby. A mother needs on average 800–1,000 extra calories per day during the end of pregnancy and breastfeeding. From an evolutionary perspective, women who easily put on weight had an advantage over those who could not, because they were able to withstand the negative consequences of lack of food or temporary starvation and still grow and feed a healthy baby.

Oxytocin contributes to pregnancy weight gain in several ways. It may influence the neuroendocrine mechanisms that control food intake, increasing hunger and decreasing satiety. This will of course result in increased eating and weight gain.

*Optimised digestion and metabolism*

A second mechanism by which oxytocin facilitates weight gain is by increased activity in the oxytocin nerves in the brain that control the function of the vagal nerve, which connects the brain with the viscera, for example the gastrointestinal tract. Increased vagal nerve activity results in an optimised digestive process and an increase in the release of insulin and hormones that support insulin release. As insulin is the hormone that helps glucose and other metabolites to be stored as glycogen and fat, this will promote the storage of energy and lead to weight gain, and in this way the use of ingested calories (food that is eaten) becomes more efficient. This increased efficiency reduces the amount of food a pregnant woman has to consume, thanks to the oxytocin-induced increased activity in the vagal nerve.

There is an even more basic effect induced by oxytocin in the gastrointestinal tract: the lowering of the levels of the hormone somatostatin. When the vagal nerve is activated by oxytocin, levels of somatostatin fall. Somatostatin is produced in the gastrointestinal tract and it inhibits the function of the entire gastrointestinal tract and the release of all the hormones that are produced there. In this way somatostatin acts as a brake on the function of the gastrointestinal tract. When levels of somatostatin are high, the function of the gastrointestinal tract is low, whereas when levels of somatostatin are low, the function of the gastrointestinal tract is high.

The importance of low levels of maternal somatostatin is shown by the fact that a baby's birth weight, and the weight of the placenta, are higher the lower the levels of maternal somatostatin are. But the system is even more sophisticated than that. If a mother is carrying twins, her somatostatin levels are lower than in mothers having just one baby. This makes sense, because a mother carrying multiple babies needs to make even better use of ingested calories to protect her energy balance than those with just one baby.

The link between oxytocin levels and somatostatin is also clear. High levels of oxytocin are associated with low somatostatin levels and higher birthweights.

*Reduced use of calories*

Mothers have yet another way to save energy during pregnancy, which is to reduce unnecessary burning or use of calories. Each day our muscles create extra heat by burning calories. Oxytocin may block this burning of extra calories during pregnancy. This effect of oxytocin is mediated via decreased activity in the sympathetic nervous system. The mother may also move less, which of course also saves calories.

All the energy-saving mechanisms we have looked at are of less importance in the western world than elsewhere, because we have plenty of food. However, these adaptations become more important in times of limited food supply. Today this rarely happens in the western world, but in developing countries lack of food – particularly high-quality food – is still a big problem for pregnant women.

*Morning sickness*

The symptoms of morning sickness may be explained by the relatively sudden and robust change in the balance of the autonomic nervous system that occurs at the beginning of pregnancy. Oxytocin (in the presence of high oestrogen levels) causes an increase in parasympathetic/vagal nerve activity and a decrease in sympathetic nervous tone.

During early pregnancy blood pressure is often very low, which is a sign of decreased sympathetic nervous tone. The increased parasympathetic/vagal nervous activity manifests as sleepiness, particularly after eating. The emptying of the stomach may be delayed, which can be experienced as sickness. Insulin release is enhanced, and storage of nutrients is promoted, to the extent that slight hypoglycaemia may be induced. A mother may have difficulty accessing energy for normal activity, because calories are being transferred to storage rather than recruited for production of energy. As a consequence of all these changes – slight hypoglycaemia, low blood pressure and changed motility in the gastrointestinal tract – a pregnant woman may feel dizzy and sick. With this perspective morning sickness is a consequence of a rapid shift to increased parasympathetic/vagal activity and decreased sympathetic nervous tone, which occurs in response to oxytocin during early pregnancy, a shift which is intended to prepare the mother for motherhood. In this way morning sickness is not an 'illness'; rather, it is a reflection of basic physiological changes taking place in early pregnancy. Some women are more sensitive to these changes than others, which explains why some women suffer terrible morning sickness while others appear to sail through early pregnancy. In some women psychological reactions become linked to the physical symptoms, which can lead to more severe debility and perhaps even hyperemesis gravidarum. It is likely that variations in levels of oestrogen and sensitivity in individual women contribute to the different ways in which women experience morning sickness. There can even be variation in the same individual: women may be affected differently in different pregnancies.

## The system of energy conservation

The term 'energy conservation system' is used to describe a number of different effects or functions in the brain, which can be activated together in order to optimise the use of energy for storage. When this pattern of effects is activated, the focus of the individual will be on food intake. Hunger is increased and digestion is optimised. In addition, the metabolism of the ingested food is directed towards storing energy, and insulin levels are high. Furthermore, stress levels decrease: levels of cortisol and activity in the sympathetic nervous system are reduced to minimise energy expenditure. Energy may also be conserved by a reduction in activity and movement.

Many of the effects of the system of energy conservation are caused by decreased function in one of the most fundamental signalling systems in the brain, the noradrenergic system. The noradrenergic system is linked to activity and aggression.

Oxytocin in the brain can activate the energy conservation system by decreasing activity in the noradrenergic system. Oxytocin induces such effects during pregnancy and to a certain extent also during breastfeeding, which we will look at later.

Pregnancy and breastfeeding are, however, not the only situations in which the energy conservation system is activated. In fact, this system is much more basic and can be triggered in many different situations to protect an individual from the consequences of insufficient food intake. From an evolutionary point of view, the most important and basic activator of this system is lack of food and starvation, which is still prevalent in many parts of the world.

*Long-term activation*

Once the energy conservation system has been activated due to a lack of food, it may continue to be activated even if food becomes plentiful again. This is probably an inborn protective system in case of future lack of food.

The system of energy conservation and its long-term effects are activated in response to any type of reduction in intake of calories, whether due to an actual shortage of food or to self-imposed restriction. Have you watched the TV programmes in which very overweight people are helped to lose enormous amounts of weight? The motivation to compete with others and to receive appreciation from viewers helps them overcome their hunger for a while and they eat very little and lose a lot of weight. The sad part of this story is that when they return to normal eating habits they will quickly gain weight again. The underlying problem is that during the intense period of weight reduction they have eaten very little and as a consequence the energy conservation system has been activated. When they reach their ideal weight and want to start eating normally again, they rapidly gain weight. The activation of the energy saving system caused by hard dieting persists and they will not be able to eat as much as they did before dieting, if they want to remain at their new weight.

*Problems with weight loss after birth*

Some women notice that they have trouble keeping their weight down after giving birth and perhaps breastfeeding. They simply can't eat as much as they used to. This is a consequence of some of the metabolic changes that occur during pregnancy and breastfeeding aimed at saving energy that are not fully reversed.

### Transfer of calories to the baby

During the second phase of pregnancy, when the baby really starts to grow, it becomes more and more important that the mother not only stores energy herself, but also that she uses energy from her own fat stores to transfer to the placenta and the growing baby. During this period blood glucose levels are higher and activity in the stress system increases. Very often mothers feel full of energy during this phase of pregnancy. Higher levels of oxytocin in the circulation facilitate the passage of energy from maternal stores to the placenta and the baby.

### Summary

- Oxytocin levels increase gradually (up to 3 or 4 times the original level) during pregnancy in response to increasing levels of oestrogen.
- Oxytocin released in the brain helps the mother to become more socially interactive during pregnancy, which allows her to communicate with and seek and receive information, support and help from her partner, parents, other members of the family and friends.

- Oxytocin also shifts the mother's interest from the surrounding world to her future baby, initiating the process of interaction and bonding with the baby.
- Oxytocin induces important metabolic changes during pregnancy, which help the mother handle ingested nutrients in an efficient way in order to allow the baby to grow.
- Early in pregnancy oxytocin helps the mother store calories and put on weight. Later on higher levels of oxytocin help the mother transfer more and more energy to the rapidly growing baby.

# 3 Oxytocin during birth

Labour usually starts after 38–40 weeks of pregnancy. We still don't know how labour is initiated in human mothers, and whether the trigger comes from the mother or the baby. Mothers know when they are in labour because of the onset of contractions, but this can be a gradual process and many mothers experience preterm contractions for a period before birth, for example at night, before the 'real' labour gets going.

## Oxytocin stimulates labour

Oxytocin plays an important role during labour because it stimulates or facilitates many different functions that are essential during birth. The ability of oxytocin to promote uterine contractions was one of the very first effects of oxytocin to be described. When Sir Henry Dale exposed small strips of uterine muscle to oxytocin, the strips contracted, which means that oxytocin acts directly on the uterine muscle to make it contract.

*Increased function of oxytocin receptors*

The number and function of oxytocin receptors (sites on muscles in the womb that recognise and bind to oxytocin) are also important in the onset of labour, as well as for the intensity of labour. The rise of oestrogen levels during pregnancy increases the function of the uterine receptors about 100-fold. This means that at the end of pregnancy much lower oxytocin levels can induce contractions of the uterus than in normal conditions. Only very small amounts of oxytocin are needed to produce contractions of the muscles in the uterus during labour.

*Oxytocin levels during labour*

When oxytocin levels are measured in the blood during labour, there is no specific time point at which oxytocin levels suddenly rise to initiate labour. As mentioned above, oxytocin levels rise to 3 or 4 times their normal level during pregnancy and near to the birth a few short-lasting peaks of oxytocin start to appear. At the beginning of labour pulses of oxytocin become more frequent, but still they occur at relatively long and irregular intervals. As labour goes on, oxytocin peaks occur more and more frequently and towards the end of labour they reach a maximal frequency of three pulses per 10 minutes. The pulses also become bigger over time. As the frequency and size of the oxytocin pulses increase, the number of uterine contractions increases and labour progresses.

*The cervix must open*

Unless the cervix is open, it is not possible for the uterine muscles to push the baby out: it is simply not possible to push something out through a closed door. During opening of the cervix, the circular muscles in the cervix must relax. The mechanisms that regulate contractions of the muscles of the uterus and the relaxation of the muscles of the cervix are partly the same, and oxytocin participates in both processes. Other substances, called prostaglandins, are also of vital importance for the process of opening of the cervix, and oxytocin contributes to the production of prostaglandins. As will be discussed later on in this chapter, the autonomic nervous system is also involved in the process of labour.

*The Fergusson reflex*

The unborn baby contributes to the release of oxytocin and to the progress of labour. In response to uterine contractions the baby's head is pushed downwards and during this movement the head puts pressure on the cervix and the upper part of the vagina. The enhanced pressure on the tissues of the cervix and vagina activates sensory nerves in these regions, which in turn trigger a release of oxytocin from the pituitary in the brain. As more oxytocin is released into the bloodstream, the uterine contractions increase and the pressure of the baby's head on the cervix and vagina becomes even stronger. In this way a positive feedback loop for oxytocin release is created. This reflex mechanism is called the Fergusson reflex.

Some researchers have collected blood samples at very short intervals during birth, and they found that a particularly big peak of oxytocin occurs when the baby is born. This oxytocin peak is three to four times bigger than previous peaks during labour, and helps the placenta to detach from the wall of the uterus and be delivered. Oxytocin released after birth is also of critical importance for another reason: by stimulating contractions of the uterus, oxytocin inhibits maternal bleeding after birth.

*The autonomic nervous system influences labour and birth*

As in the control of the function of most other organs, the

autonomic nervous system plays an important role in regulating the function of the uterus. We often forget that the autonomic nervous system participates in the control of uterine contractions and thereby the progress of labour. In fact, rats don't need any oxytocin at all to give birth; the autonomic nervous system can do it all by itself. But under normal circumstances circulating oxytocin and the autonomic innervation of the uterus and the cervix act together and both contribute to the regulation of the process of labour and birth.

The parasympathetic branch of the autonomic nervous system, which stimulates processes that are related to growth, restoration and reproduction, plays an important role during birth. It helps oxytocin induce uterine contractions and also promotes blood flow in the uterus and thereby in the placenta. There is a very close link between oxytocin and the parasympathetic nerves that innervate the uterus. In fact, oxytocin nerves, which have their origin in the PVN within the hypothalamus of the brain (see Chapter 2), reach down to the very end of the spinal cord and connect to the parasympathetic innervation of the uterus. So oxytocin may influence the contractions of the uterus both via the bloodstream and via the parasympathetic nerves.

The sympathetic nervous system, on the other hand, is linked to physical activity and even stress responses. Too much activity in the sympathetic nerves during labour tends to induce different uterine contractions, which are long and painful. In addition, it decreases blood flow in the uterus and placenta, which is undesirable during labour. The relaxation of the muscles in the cervix will also be delayed if the sympathetic nervous system is too active.

When stress levels are very high, adrenaline may be released from the adrenal gland and reach the uterus via the bloodstream. In this situation high levels of adrenaline can cause labour to stop completely.

In general, the higher the parasympathetic nervous tone, the better the progress of labour.

## Oxytocin, the shy hormone

Subtle environmental cues, such as an unfamiliar environment or the presence of a strange person, can stop the release of oxytocin and inhibit the start or progress of labour. This oxytocin brake is often activated during the very early stages of labour.

Historically, women gave birth without doctors and hospitals, often in their homes. Even earlier than that, women probably chose to give birth in a sheltered place, such as their own cave, or another safe place, because giving birth can be dangerous, not only because of the process of giving birth itself, but because of environmental factors. When a woman is giving birth she can't protect herself (and her newborn) from potential predators, so she needs shelter and protection.

### *The oxytocin brake*

To make sure that women didn't give birth unless they were in a safe place, evolution built a brake into the oxytocin system. If a woman who was going to give birth experienced her surroundings as threatening or even just unfamiliar, oxytocin simply wasn't released. This gave the mother time to go home or at least to find shelter, protection and support. You can turn this statement around and say that the release of oxytocin in connection with labour 'prefers' a safe and familiar environment. This early oxytocin brake is not identical to the arrest of labour caused by intense stress and high levels of adrenaline later on in labour. The mother may not feel at all stressed, and labour is merely postponed rather than stopped.

*True versus false safety*

Today the majority of women give birth in a hospital. To give birth in a hospital might be considered to be 'super safe', with all the expert midwives and doctors present to take care of you and your baby. The problem is that the birthing mothers of today still have the archaic instinctual reactions to feelings of safety or unsafety, or even familiarity versus unfamiliarity, that kept their ancestors safe. The modern feeling of safety based on expertise doesn't fit into this historic pattern, and may be experienced as false by the mother's unconscious mind checking her surroundings. In fact, hospital may be experienced or misinterpreted as an unfamiliar and unsafe place, which activates the oxytocin brake and prevents the further progress of labour. Many women have experienced how regular contractions have started at home, in particular during the night, and how these contractions made them hurry to the labour ward, only to find that when they arrived there the contractions stopped.

When labour stops some mothers are actually sent home from the labour ward and asked to come back when labour starts again. Others receive medical interventions to speed labour up, as midwives and doctors may interpret the absence of contractions as a failure of the mother's own physiological capacity to give birth. Women may be given epidural analgesia to relieve pain, or an oxytocin infusion to increase the contractions of the uterus, or both these interventions. Epidural analgesia reduces oxytocin release, causing contractions to become weaker, meaning that the mother might be given an infusion of oxytocin to compensate. Alternatively, if a woman is given an infusion of oxytocin, which is often linked to painful contractions, she may need an epidural to alleviate the pain. When this happens the birthing woman may easily get the feeling that she can't give birth herself and lose trust in her own competence.

## How support can help mothers giving birth

In addition to wanting to give birth in a safe and familiar place, women giving birth often want company. They want to have somebody around in order not to be alone, and also to have someone to support and help them when needed. Before the era of hospital births this supportive role was often played by female relatives or other wise and experienced women.

The basic and important need that women giving birth have for the support of another person has only recently been fully acknowledged. The role of midwives has been revised and extended so that not only do they help women with the practical aspects of give birth, but they also support them physically and mentally during labour and birth. Nowadays a person who specialises in helping mothers giving birth, a doula, may also be present during birth. Of course the presence of the father at the birth may serve the same purpose and help mothers give birth; but doulas are able to extend the support they offer to the father as well.

### *Quicker and more positive birth*

Many scientific studies show that birth is faster and less painful in the presence of a skilled supportive person. The need for medical interventions, such as caesarean sections, epidural analgesia and infusions of synthetic oxytocin is also reduced. Studies also suggest that mothers are more likely to feel well, happy and content after birth if they received support during birth. They feel proud and their self-confidence increases. Positive feelings they have about their newborn and even their partner may become stronger.

The big question is, of course, by what mechanisms does the presence of a supportive woman during birth and labour induce these miraculous effects? How is the progress of labour enhanced? How are the mother's feelings influenced in such a positive way? The answer is simple: the presence of a supportive woman stimulates the release of oxytocin into the circulation and into the brain. This enhances the activity in the parasympathetic nerves and decreases the activity in the sympathetic nervous system.

How touch, warm hands and stroking can reduce stress levels and pain and stimulate wellbeing will be described in more detail in the chapter on skin-to-skin contact. To give love and support is, however, also an active mental process. By being calm, open, sensitive, supportive and in a loving mood, the helping person may make the woman giving birth less fearful and more open and trusting. When the birthing woman feels calm and without fear, the oxytocin system operates at a maximal capacity and so does the parasympathetic nervous system. In addition, the stress-reducing effects of touch and warm hands will be optimised. So the doula can be thought of as enhancing the power of the birthing mother's oxytocin system.

## Oxytocin and the brain

As described in Chapter 2, oxytocin not only acts via the circulation, but is also released from nerves within the brain to influence functions in the brain itself.

Many studies show that substantial changes occur in the morphology and function of the maternal brain, some of which are linked to the effects of rising oestrogen and oxytocin levels during pregnancy. The number of receptors for these hormones also increases. Indeed, oxytocin levels in the brain increase during labour in human mothers.

Animal experiments demonstrate that oxytocin is released in parallel into the circulation and into the brain during labour and birth. As also mentioned in Chapter 2, oxytocin induces an inborn programme called 'maternal behaviour' in all mammalian mothers, which enables them to intuitively know how to care for, interact with and protect their offspring after birth. These behaviours are induced by oxytocin released into the brain during labour and birth, in part as a response to the Fergusson reflex. Experiments in sheep have shown that if oxytocin release is blocked in some way during labour, for example by a type of pain relief called peridural analgesia, ewes don't express any interest in their newborns, don't care for them and don't bond with them. Fortunately, this black and white scenario does not occur in the same way in human mothers. But as we will see later in this chapter, there are some psychological adaptations in human mothers that might represent some remnants of an inborn oxytocin-linked 'maternal behaviour' programme.

*Humans and inborn reactions*

The human brain is different from those of other mammals and our behaviour is less governed by instinctual reactions. We learn how to behave and perform from parents, friends, newspapers and the internet. But we do have some remnants of some innate archaic mammalian reaction patterns and behaviours, which are activated during pregnancy, labour and birth, without us knowing it. It would be interesting to know if human mothers would know how to handle a baby if they for some reason were alone and hadn't received all the information that we automatically obtain today. I bet they would!

### Pain-relieving effect

One important effect caused by oxytocin in the mother's brain during labour is a decreased experience of pain. Oxytocin decreases pain by several mechanisms. There are nerves containing oxytocin that reach all the way down in the spinal cord to inhibit the activity of incoming nerve fibres mediating pain, and in addition there are several areas in the brain itself that are involved in the perception of pain and which are innervated by oxytocin nerves. In other words, mothers have their own pain-relief system that operates during labour and birth and which is activated by labour itself, since labour contractions are associated with a release of oxytocin, not only into the circulation, but also the brain. Oxytocin release is increased by the Fergusson reflex, which is activated by the pressure of the unborn baby's head on the cervix and the upper part of the vagina. When oxytocin is released in the brain, mechanisms that are linked to pain relief are activated.

The experience of pain during birth can be very variable. Some women have easier, less painful births, while for others it is more difficult. It is obvious that the birthing woman's state of mind matters. If she feels calm and relaxed, feels secure and is surrounded by people that she knows well or otherwise trusts, the experience of birth is more positive, and pain is less pronounced. It is also obvious that stress levels are lower and birth is quicker in this situtation. Many women who have had a home birth describe their birth in more positive terms and report less pain than those who have given birth in medical settings.

*Amnesic effect*

There is an additional effect of oxytocin that is related to the pain-relieving effect: the amnesic effect. Mothers tend to forget the difficult pain they endured during birth. Mothers who have given birth score the pain of labour and birth as very intense and hardly bearable immediately after birth, but after some time has passed they rate the pain as less difficult and more tolerable.

Through its amnesic effect oxytocin actually changes the memory of the difficult experience of giving birth to a less disturbing and frightening one. This is nature's way of helping to ensure that mothers are likely to be able to give birth again in future, without being affected by the memory of pain they experienced. It is only when mothers actually give birth again that they remember how intense the pain was the first time!

### Preparation for interaction and bonding

*Stimulation of the reward centre towards the end of labour*

There is most likely a substantial increase in oxytocin levels in the brain towards the end of birth, just as in the circulation. It is common that women who have had a natural vaginal birth experience very strong feelings of happiness and joy right after birth. These positive feelings are of course a consequence of the situation itself, of having experienced the miracle of having given birth to a baby. There might, in addition, be a neuroendocrine explanation for this experience, as oxytocin may activate the reward system in the brain leading to higher levels of dopamine. In this way the positive, sometimes euphoric feelings that mothers experience when a baby is born may be boosted by the action of oxytocin in the brain. The process of bonding, or attachment between mother and infant, will be dealt with in more detail in the next chapter, in which we will look at the consequences of skin-to-skin contact after birth.

*The maternal personality adapts to motherhood*

Mothers who have had a normal physiological birth are on average more interested in social interactions, including being somewhat pleasing and wishing to conform with others, in comparison to control women of the same age. The increased preference for social interaction of course makes it easier for the mother to ask for and receive help from other women during labour and birth. In addition, it facilitates her future interaction and bonding with her newborn. These adaptations are likely to represent some small human remnants of the more instinctual maternal behaviour that comes into play after birth in other mammalian species.

An increasing number of observations show that mothers who have not had a natural vaginal birth, but instead a birth with different types of medical interventions, may behave and experience themselves slightly differently after birth. It is in fact possible that the brains of these mothers were not flooded with oxytocin during birth to the same extent as those mothers who had a normal vaginal birth. These effects will be discussed in the chapters on medical interventions during birth.

*The sacredness of birth*

Giving birth can be a very positive experience if everything goes well. It can also be very empowering, because it may give the mother a sense of having achieved something important. Perhaps the most important thing in life. There are many women who try to bring back or recapture this glorious image of birth, instead of seeing birth as something necessary that women just have to go through in order to have a baby, and which you can make almost unnoticeable by having an elective caesarean or interventions that reduce your pain. These women try to bring forward the picture of another kind of birth, controlled by the birthing woman herself and not by the medical system. They want births to be the happiest and most joyous moments of life, serving as a transition rite, making women prouder and more conscious of their strength and power than ever before.

### Negative experiences of birth

Sometimes the natural pain and stress-relieving systems that should help mothers with the pain and fear in connection with birth don't work or are insufficient. The birth may be complicated for different reasons, or the natural protection against pain and stress, including oxytocin release, may not be sufficiently activated.

*Strong pain and fear inhibit oxytocin*

The brain is, of course, continuously informed by sensory nerves from the womb that mediate the sensation of pain. If the mother experiences very intense pain during birth she might become frightened and her stress levels will increase in response to such feelings.

Fear, pain and high stress levels are no good for labour. Oxytocin release is inhibited and the sympathetic nerves innervating the uterus may be activated, so labour does not progress. This is why medical interventions such as epidural analgesia have been developed. In a situation like the one described above, the epidural may restart the progress of labour by reducing pain, fear and stress.

Still the problem may not be completely solved. If a mother has had a very tough time during labour, the memory of this negative experience might get 'stuck' and her stress system might become a little bit overactive. If this state continues the mother may develop negative psychological reactions after birth. In these situations, it is not uncommon for mothers to develop anxiety and depression after birth, or even post-traumatic stress disorder (PTSD). All of these conditions, which may be very difficult for the mother to endure and which may disturb the relationship between her and her baby, are linked to a sustained elevation of activity in the stress system.

*Other types of pain relief*

Obviously everything should be done to help mothers with unbearable pain during labour. It seems that the presence of a friendly supportive person during birth, or, as will be discussed in the next chapter, skin-to-skin contact with the baby after birth, may have a protective effect against not only pain during labour, but also the risk of developing anxiety and PTSD. The reason for this is that when the function of the maternal oxytocin system is stimulated, the activity in the HPA axis and the sympathetic nervous system and stress reactions and memories of pain will be reduced. With optimal activation of the oxytocin system, the beneficial amnesic effects of oxytocin will be activated.

### The baby

Much less is known about the function of oxytocin during labour and birth in the baby than is known about the function of oxytocin in the mother. Many people don't even know that the baby produces its own oxytocin. They assume that the baby receives its oxytocin from the mother, via the umbilical cord. However, oxytocin levels in the baby are at least as high or even higher than those of the mother right after birth, irrespective of the type of birth, which supports the idea that the baby produces its own oxytocin and is not just a passive receiver of the mother's oxytocin.

The production of oxytocin starts very early during embryonic or foetal life. At the beginning of life oxytocin is probably produced in foetal cells and then secreted into the surrounding tissues to influence the function of neighbouring cells. Later on, when the baby is more developed, oxytocin will be produced in the hypothalamus and secreted from the pituitary into the circulation and from nerves within the brain to influence brain function. In this way oxytocin will influence the development of important behavioural (for example, social interactive behaviours) and physiological functions in the baby before birth.

*The baby's oxytocin levels rise during birth*

Oxytocin is released within the baby in connection with birth, and oxytocin levels are higher in the baby after a birth by vaginal delivery, compared to babies born by elective caesarean section. Mothers having an elective, planned or preterm caesarean don't experience labour with oxytocin-induced contractions, so their infants are not exposed to muscular contractions that would trigger the baby to release oxytocin. But is this the whole story?

*What triggers oxytocin release in the baby during birth?*

As mentioned previously, oxytocin is released in response to pleasant stimulation of the skin, and this subject will be discussed in more detail in the chapter on skin-to-skin contact. Oxytocin is, however, also sometimes released in response to painful stimulation, to reduce pain and other consequences of stressful situations.

*The baby also perceives pain during birth and oxytocin reduces it*

During labour the baby is exposed to severe stress and most likely also to pain. We tend to believe that birth is a situation which is linked to pain only in the mother. But why should pain during birth be restricted to the mother? Uterine contractions give rise to pain in the mother, and it is likely that intense contractions also give rise to pain in the baby during labour, and in particular during birth. The pain-conducting nerve fibres that originate in the skin and other tissues of the body of the baby could be activated during birth. The pressure induced by contractions is likely to activate pain receptors in the skin, and the baby's head is subject to severe squeezing and high pressure during birth, which could be linked to pain.

Luckily, oxytocin released in the baby's brain helps to reduce the baby's experience of the pain of being born.

*Oxytocin reduces damage caused by low levels of oxygen*
During labour, in particular if the contractions are strong and occur with a very high frequency, the delivery of blood from the placenta to the baby may be a little compromised. Under these circumstances the baby may receive less oxygen and nutrients from the mother. In such stressful conditions, the pH of the baby's blood may fall, and levels of the stress hormone adrenaline may rise to counteract the problem. Interestingly, the baby's oxytocin levels are also higher under these stressful conditions and oxytocin is released in the baby during birth in direct response to the severity of the situation. This is where the anti-inflammatory effects of oxytocin come in to play. When the baby is exposed to periods of low levels of oxygen, slight damage might be induced. Oxytocin may protect the baby from such negative consequences by reducing inflammation and stimulating the growth of new cells. In this way the extent of tissue damage can be substantially reduced by oxytocin produced by and secreted in the baby.

*Does oxytocin diminish the trauma of being born?*
Do you remember that oxytocin released in the brain of the mother not only decreases the experience of pain during labour, but in addition weakens the memory of the pain of birth via the amnesic effect? There is every reason to assume that the newborn baby also experiences these sustained protective effects against the memory of stress and pain during labour and birth from its own release of oxytocin.

The trauma of birth is a well-recognised term in the field of psychology. The birth trauma of the baby probably has the same roots as that of the mother: a sustained memory or experience of the very intense pain, fear and stress induced during birth. In the baby, however, the experience of hypoxia, including anxiety and perhaps hypoxic pain as a consequence of low levels of oxygen, might also contribute to such negative experiences. Therefore, oxytocin released during birth may well, due to its amnesic effect, protect the baby from long-term negative consequences of the experience of unbearable stress and pain during birth.

*Preparation for social interaction and bonding with the mother*
Oxytocin released during birth also prepares the baby for interaction and bonding with the mother. These effects induced by oxytocin during birth will be further stimulated after birth, in particular if the baby is allowed to be in skin-to-skin contact with the mother (or the father). These effects are described in more detail in the next chapter, which looks at the potent behavioural and physiological effects of skin-to-skin contact after birth in both mother and newborn.

## Summary

The mother:

- Oxytocin in the circulation contracts the muscles of the uterus and relaxes the muscles of the cervix, thereby stimulating the progress of labour and birth.
- Oxytocin is released in short pulses with increasing frequency and size during labour and birth. The biggest rise of oxytocin occurs in connection with birth.
- The parasympathetic nerves enhance and the sympathetic nerves counteract the effects of oxytocin and the birth process.
- Fear, stress and pain inhibit the birth process, whereas friendly support, touch and warmth stimulate it.
- Oxytocin release and the progress of labour is facilitated when mothers feel safe and relaxed, for example when they are in a familiar environment and surrounded by people they know and trust. Mild stressors such as an unfamiliar

environment or the presence of an unknown person may inhibit oxytocin release and thereby the initiation or progress of birth.
- Oxytocin released within the brain decreases the sensation/perception of pain and induces amnesia.
- Oxytocin reduces anxiety and stress levels.
- Oxytocin prepares the mother's brain for motherhood by making her more open to social interaction.
- Oxytocin may also stimulate the reward centre in the brain and cause dopamine release, which helps mothers experience joy and happiness.
- Sometimes birth may be a very negative experience linked to intense fear, stress and pain. As a consequence of this some mothers develop anxiety, depression or even PTSD after birth.

The baby:
- Oxytocin is produced by and released into the circulation and the brain of the baby.
- Oxytocin may support the growth and development of the baby.
- Oxytocin may reduce pain and protect the baby from damage caused by low levels of oxygen.
- Oxytocin may help the baby forget the pain and fear induced during labour and birth through its amnesic effects.
- Oxytocin may prepare the baby for interaction and bonding with the mother.

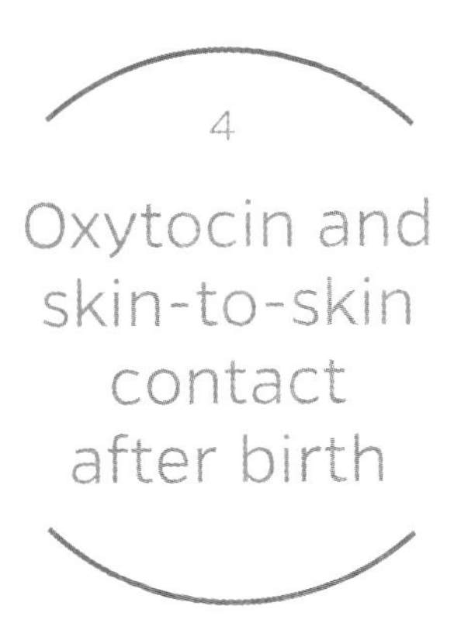

# 4

# Oxytocin and skin-to-skin contact after birth

## The mother and the newborn

### *The practice of skin-to-skin contact*

If you have previously given birth, your newborn may have been placed on your chest immediately after birth, and you may have experienced how positive that was. This clinical practice is called 'skin-to-skin' and means that a midwife, or another member of staff on the labour ward, puts the naked newborn baby on the mother's or father's chest more or less immediately after birth. Babies are most often put on the mother's chest, but if the mother has given birth by caesarean or has other complications, the baby's father may be offered skin-to-skin instead.

There are so many reasons why this practice is now implemented more and more often in modern maternity care. It activates social interactive behaviours in mothers and newborns, which are beneficial for the mother and the baby and their future relationship. It also activates important adaptive physiological reactions. It may decrease the risk of hypothermia in the newborn and increase the baby's skin temperature. It also helps increase the baby's glucose and oxygen levels, helps decrease and regularise the newborn's heart rate and decreases cortisol levels. The mother's cortisol levels also decrease, as well as markers for inflammation. Her chest skin temperature increases and some vagally controlled functions in the gastrointestinal tract are activated. Together all these effects promote wellbeing and health in both mothers and infants, not only in the short but also in the long term. This will be discussed more in detail later on.

*To be born is to be separated from a warm and pleasant environment*

When the baby is within the uterus, it is surrounded by amniotic fluid. This is experienced by the baby as a feeling of warmth and light pressure on the skin, which will of course give rise to a pleasant awareness of the contours of the body. When the baby is born it suddenly and abruptly loses this feeling of pleasant contact with the surrounding world. The baby cries after birth, in part because of the loss of the feeling of warmth and light supportive pressure previously provided by the mother via the amniotic fluid. In fact, when contact with the amniotic fluid ceases, the newborn experiences a kind of loss, or separation anxiety. That's why the baby immediately stops crying if it is put on the mother's chest after birth. The skin-to-skin contact provides the newborn with a renewed contact with the mother, but this time it is via light pressure and warmth on its front side, from lying on the mother's chest.

*Mothers may serve as incubators*

In animals in which the offspring are born very immature, continued close contact between mother and offspring after birth is necessary for survival. The mother's closeness, her warmth, and the sounds and odours that she transmits to her offspring, regulate their behaviours and physiological functions. Skin-to-skin contact between mother or father and a newborn human baby serves the same purpose, as it influences and stimulates the behaviour and physiological functions of the newborn. But in humans the interaction is bilateral, as the mothers are also influenced by the close contact.

### Giving of warmth

One important mechanism by which the mother influences the newborn via skin-to-skin is by providing warmth. The skin of the mother's chest radiates heat and in this way provides the baby with energy. In many ways giving warmth is as important as giving milk. When mothers warm their newborns they resemble birds incubating eggs. When birds incubate their eggs, the eggs warm up and warmth is transferred through the eggshell to the baby bird inside the eggs, which helps them grow and finally to hatch.

#### *Flushing of the skin*

During birth and afterwards, the skin of the mother's face and chest starts to flush. This reddening of the skin occurs because blood flow in the blood vessels of the skin of the face and chest increases and therefore skin temperature rises. The increase in chest temperature occurs in pulses. Oxytocin released in the mother during skin-to-skin contact with the newborn is responsible for the pulsatile increase in chest skin temperature. No such increase occurs in mothers that are separated from their newborns, showing that the presence of the baby triggers the temperature variations.

The warmth babies receive from their mothers during the skin-to-skin period helps them increase their own temperature and prevents them from dangerous hypothermia – getting too cold – which may happen after birth when the baby is no longer in the warm uterus. At the same time the newborn's access to energy increases as their blood glucose levels go up and their interaction with the mother is stimulated.

*Increased skin temperature in the infant*

When a baby receives warmth from its mother, the temperature of the skin increases. So newborns mirror their mothers by increasing the blood flow in the blood vessels of their skin. This effect is most easily observed in peripheral parts of the body, such as the hands and feet, but in fact all areas of the skin are affected. An observant mother or midwife can see how the feet or hands of the newborn baby turn pink within a few minutes after birth provided the baby is allowed to be in skin-to-skin contact. This change of colour, which is caused by an increased blood flow to the skin, does not occur if the infant is not in close contact with the mother after birth.

During skin-to skin contact mother and baby even synchronise their skin temperature, as if they want to continue to be one individual, as they were during pregnancy when the baby was still in the mother's womb. During this period the baby's skin temperature approaches that of the mother. The warmer the mother, the more the baby´s skin temperature rises.

## The newborn baby approaches the mother

*Remnants of mammalian clinging behaviour*

I am sure you have seen pictures of young monkeys and apes clinging to their mothers. Monkey babies sometimes cling to their mothers' backs, whereas babies of the more developed apes cling to their mothers' chests. Human babies are a bit immature when they are born and can't perform clinging behaviours by themselves like great big ape babies. In ancient times human mothers most likely picked their newborns up, looked at them and then brought them to their chests, not necessarily only to breastfeed, but also to warm them and to hold and protect them.

*Babies' crawling behaviour*

Human babies can still initiate interaction with their mothers. If a newborn baby is put on its mother's chest immediately after birth, and the mother is asked not to interfere, the newborn will start to perform a sequence of activities aimed at approaching the mother's breast. The baby interacts with the mother in different ways and slowly crawls towards the mother's breast, governed by smell and vision, and starts to breastfeed before falling asleep. The whole sequence of events takes about 90 minutes. As the baby makes its way to the breast it performs crawling movements, massage-like movements on the mother's breasts, and licks and touches the nipple. The baby's ability to crawl and breastfeed often makes mothers feel very proud and happy. Perhaps they also feel a little relieved, because the baby knows what to do.

This is not really a natural situation, because in normal circumstances most mothers will help their babies to start to breastfeed by holding them and helping them with their hands. The newborn's crawling behaviour is nonetheless interesting, because it shows that humans share a lot of instinctual behaviours with more primitive animals. The crawling behaviour of the newborn baby is a little like the kangaroo baby, which crawls upwards inside its mother's pouch to find the mother's nipple.

Obviously skin-to-skin contact after birth is not essential for survival, since many babies have survived and thrived without it, but as we can see it may stimulate positive behaviours and physiological reactions in both mother and baby, which may have long-term positive consequences for their relationship and perhaps also for their relationships in general and their health.

*Mother's warmth triggers the newborn's crawling*

The warmth given to a newborn from the mother's chest is one important trigger of the approach responses. The newborn becomes energised by the warmth they receive from their mother, and the crawling or approach behaviour becomes activated. This effect is most likely caused by oxytocin released within the baby in response to the warmth it has received from the mother. In support of this and as will be discussed later on, newborns of mothers who have received extra synthetic oxytocin during labour interact more with their mothers during skin-to-skin contact after birth. They perform more crawling and also licking and touching of the mother's breast in order to become familiar with the mother. Most certainly these mothers have transmitted more warmth to their newborns.

*Release of oxytocin in the mother*

While crawling on their mother's chest, babies perform massage-like movements, similar to those of kittens before suckling. These movements are not so easy to see at normal speed, but become obvious if the crawling behaviour is filmed. The massage-like movements are important activators in the interaction between mother and newborn, because they increase the mother's release of oxytocin. In fact, the effect on the mother's oxytocin release is dose-dependent: the more the baby massages the mother's breast with their hands, the more oxytocin will be released by the mother.

*Mothers interact more*

When oxytocin is released in response to skin-to-skin contact in the mother, this is not only in the circulation, but also in the brain, which is flooded with oxytocin. Of course this increase in maternal oxytocin levels results in more maternal care and interaction with the baby, and it increases the mother's sensitivity to the newborn's needs. In this way a positive loop is created. It's as if the crawling newborn is not only asking its mother for milk, but also for warmth, care and love.

Studies show that mothers' approach to and interaction with their newborns is increased by skin-to-skin contact. Mothers who are in skin-to-skin contact with their newborns touch them more, speak with a softer voice, and synchronise their touch and sounds with those of the baby more than mothers who do not have skin-to-skin contact.

In skin-to-skin contact the newborn baby is able to relax, as if they know that they are secure and safe. The increased skin temperature is a sign of this. Also the baby's levels of anxiety decrease, so that they become less fearful and dare to approach the mother. As we will discuss later, the baby's feeling of relaxation and sense of security at this point may be reflected in long-term effects that can be observed much later.

## Skin-to-skin contact reverses stress levels and promotes growth

### *Immediate decrease in stress levels in response to skin-to-skin contact*

Labour and birth are extremely stressful and energy-intensive processes. In fact, it is not possible to have a vaginal delivery or be born that way without an activated stress system. It is very stressful for the mother to push the baby out, and stressful for the baby to experience the process of being squeezed out of the womb to be born. The baby also actively participates in its birth. Thus stress levels are very high in both mother and newborn after birth. These high stress levels are of importance for the maturation of some functions in the baby, such as that of the lungs. As soon as the baby is born there is no longer any need for such high stress levels, and skin-to-skin contact after birth is linked to rapid stress-reducing effects.

For example, levels of the stress hormone cortisol fall in both mother and newborn. Also the baby's pulse rate becomes more regular. The baby's increase in skin temperature is also a sign of reduced stress levels, as blood flow to the skin is inhibited by stress. All of these effects could be linked to the release of oxytocin that occurs in response to skin-to-skin contact in both mothers and newborn babies, which in turn leads to a decrease in the activity of both the HPA axis and the sympathetic nervous system.

*Skin-to-skin contact may reduce long-term stress reactions caused by 'the trauma of giving birth' and 'the trauma of being born'*

Skin-to-skin contact after birth may also reverse the 'trauma of giving birth' in the mother and 'the trauma of birth' in a deeper sense.

As described in the previous chapter, it is not uncommon for mothers who experience unbearable pain or fear during labour and birth to develop anxiety and depression after birth, or even post-traumatic stress syndrome (PTSD). All of these conditions, which may be very difficult for the mother to endure and which may disturb the relationship between mother and baby, are linked to elevated activity in the stress system, which is expressed, for example, by high levels of cortisol.

'The trauma of birth' is a concept that is widely used within the field of psychology. Some people seem to be negatively affected by their own pain or fearful birth experiences as adults and suffer from this stress experience in many ways.

It is possible that the release of oxytocin and the decreased activity in the stress system of both the mother and the baby after exposure to skin-to-skin contact may help prevent such long-term stress-related memories, and thereby reduce the risk of developing the trauma of giving birth and the trauma of being born.

*Growth-promoting effects of skin-to-skin contact*

As mentioned above, oxytocin is released immediately after birth during skin-to-skin contact both in the mother and in the newborn baby. This is the time when the powerful anti-stress effects of oxytocin are activated in order to reduce the stress of giving birth and being born.

Now energy is needed for milk production in the mother and growth in the baby. The reduction of stress levels helps with this and, in addition, skin-to-skin contact via oxytocin release stimulates processes that are related to growth in both mother and baby. In this way the mother's milk production is stimulated and the baby's growth is more efficiently stimulated.

Babies allowed to have skin-to-skin contact after birth make better use of ingested calories than those who do not experience it. This effect is in part mediated by the powerful stimulatory effect of oxytocin on the vagal nerve, which is the big nerve (part of the parasympathetic part of the autonomic nervous system) that regulates function in the gastrointestinal tract by activating the release of hormones, including insulin, which optimise digestive processes and energy storage.

However, the mother's digestive capacity and metabolism are also affected by skin-to-skin contact. In a reciprocal way her own vagal nerve activity is stimulated by skin-to-skin and she becomes more energy efficient, even after the baby has been born and her pregnancy adaptations begin to decrease (see Chapter 1). These processes, which include elevated insulin, and which are activated by skin-to-skin contact in both baby and mother after birth, can also be activated later on by breastfeeding in the mother, and by suckling in the baby.

*Kangaroo care*

The growth-promoting effect is more clearly seen in premature and low birth weight babies who repeatedly experience skin-to-skin contact. These babies gain weight and grow more quickly if they are allowed to stay on their mother's or father's chest than if they are treated in an incubator. 'Kangaroo care', as it has become known, and which is becoming more common in special care baby units, prioritises skin-to-skin contact and consists of repeated periods of skin-to-skin. These babies not only put on weight more quickly, but they also grow and develop better. Again the effect is reciprocal and the mother's production of milk is also increased.

*Pleasant stimulation of the skin acts by releasing oxytocin and increasing parasympathetic/vagal nerve activity*

It also seems that other types of pleasant sensory stimulation of the skin work to induce the same effects. Small babies that are placed on lamb's wool gain more weight than those who lie on ordinary cotton sheets. Some types of soft massage can also help babies grow. Even the sense of 'touch' in the mucosa of the mouth provided by a pacifier or dummy works, in the sense that it stimulates increased weight gain in infants that are having to be tube-fed.

All these types of pleasant sensory stimulation activate oxytocin release, which in addition to causing a decrease in the activity of the HPA axis and the sympathetic nervous system, also increase the activity in the parasympathetic nervous system. All the positive effects on weight gain and growth caused by different types of 'touch', or stimulation of sensory nerves, are caused by oxytocin and an increased activity in the parasympathetic and vagal nerves.

*Being touched and held, a basic need*

To be touched and to be close to other people is of vital importance for life. In fact, touch and closeness is as vital for life as food. Ashley Montagu coined the term 'skin hunger' to express how vital touch is for life. Food hunger is the need for food, and ingestion of food gives rise to satiety, but also to a feeling of calm and relaxation. Skin hunger is the need for touch, a hunger that is stilled by closeness and touch and which leads to calm and relaxation. In fact, without touch and closeness infants may die, even if they receive food and their hygienic needs are fulfilled. This was observed in connection with the world wars and in the notorious Romanian orphanages that were discovered in the early 1990s. Babies simply fail to grow, and their immune systems cannot fight off infections.

The amount of touch that is needed is surprisingly low. Reports from orphanages show that infants lying closest to the door grew and developed better in comparison to infants further away. The reason turned out to be that night nurses held the baby lying closest to the door more often than the other children. This small intervention made a difference.

The symptoms of neglect and lack of touch, including apathy and lack of growth and development, are not due to any real physical damage, but are caused by the activation of physiological brakes in response to lack of stimulation and appropriate care. Therefore, if the environment changes and affected infants start to receive love and affection, including touch and closeness, they may resume growth and development.

## The role of sensory nerves during skin-to-skin contact

Now that we have established that skin-to-skin contact after birth induces many positive effects in both mother and baby, it is time to look at how these effects are brought about. It is all about stimulation of our senses. Many different types of communication are used during skin-to-skin contact. Seeing her baby, and in particular having eye contact with her newborn, is important for the mother. Smelling and hearing the baby also influences the mother in a positive direction. The newborn baby cannot see very well, but responds to the mother's eye contact and voice. The newborn also uses vision, as well as the sense of smell, to direct its crawling movements towards the mother's breast. Odours are also important cues for recognition and bonding between the mother and the baby. But there is another important source of information that is often forgotten, and that is the skin.

*Tactile interaction*

When the baby is put on the mother's chest, both mother and baby are involved in a bilateral 'skin' interaction involving warmth, touch and light pressure (the baby weighs around 3–4kg). In addition, while the newborn is crawling both mother and baby engage in a kind of stroking. All these interactions activate sensory nerves from the skin in both mother and baby.

Of course mothers also touch other areas of the baby's body; the mother may touch the baby's fingers and toes, or stroke the baby's head and back. Also in these situations sensory nerves in different parts of the baby's skin will be stimulated.

Further, sensory nerves in the skin of the palms of the mother's hands will be actively involved. Studies have investigated the effects on young boys and elderly people of stroking dogs. They showed a beautiful correlation between the amount of stroking given to the animal, and a decrease in blood pressure, heart rate, cortisol levels, and increase in finger skin temperature in the humans. These results show how important the skin is for our regulation of stress levels.

*Nerve fibres mediating painful or pleasant stimuli link to stress and anti-stress responses*

Painful or dangerous stimulation will activate the HPA axis and the sympathetic branch of the autonomic nervous system, whereas mild and pleasant stimulation will activate the oxytocin system and parasympathetic branch of the autonomic nervous system.

*Which nerve fibres are involved in skin-to-skin contact?*

Mild stimulation of nerves in animal skin demonstrates that stress levels (blood pressure, cortisol and adrenaline levels) decrease rapidly and substantially after initiation of stimulation, suggesting that activity in the HPA axis and the sympathetic nervous system is decreased. In addition, expressions of activation of the vagal nerve, such as changed levels of hormones in the gastrointestinal tract, are observed.

However, there is still no absolute consensus about which sensory nerve fibres are activated in response to skin-to-skin contact. A type of C-fibre known as Ct-fibres have been shown to be activated in response to light stroking in humans, leading to areas in the brain linked to feelings of wellbeing being activated as well. These Ct-fibres are definitely involved in the effects caused by skin-to-skin contact between mothers and babies.

*Link to oxytocin*

Stimulation of nerves in the skin is also followed by a release of oxytocin. As oxytocin in the brain decreases stress levels by reducing activity in the HPA axis and the sympathetic nervous system, it also activates of the vagal nerve and digestion and metabolism. The stress-reducing and growth-promoting effects of skin-to-skin contact, which are linked to oxytocin release, may be mediated by other types of thin C-fibres or even thicker myelinated nerve fibres.

### Other practices involving skin-to-skin contact

Skin-to-skin contact should not be thought of as a 'treatment' that takes place only at the moment of birth. There are many other situations in which closeness has been shown to exert similar positive effect on behaviour and physiology.

*Repeated treatment with skin-to-skin contact*

As the Canadian psychologist Anne Bigelow has shown, skin-to-skin contact works as a kind of treatment even after birth. If full-term babies and their mothers have some hours of extra skin-to-skin contact during the first weeks after birth, both mothers' and babies' stress levels were decreased and their interaction with each other improved. Remnants of these effects have been shown to persist as long as nine years later. These interesting data really show how crucial touch and closeness early in life are for humans.

*Kangaroo care*

Kangaroo care (skin-to-skin treatment of premature infants, as mentioned above) is based on the same principle, that closeness between the premature baby and its parents initiates physiological processes that help the infant grow and develop more quickly. Kangaroo care influences the mother in a positive way too, in the sense that she produces more milk and feels happier and more content with her role as a mother, even in the difficult circumstances of having a premature baby in special care.

*Co-sleeping*

When mothers and babies sleep together (co-sleeping), similar positive effects are exerted. Indeed, several reports show that co-sleeping is beneficial and mothers even intuitively know how to lie around their babies in order to protect them.

Unfortunately, co-sleeping has had a bad reputation, and has even been forbidden in some countries, because of accidents in which babies have been suffocated. These accidents are often linked to use of alcohol and other drugs, or unsafe co-sleeping practices. For much more information on safe co-sleeping, and the research into bedsharing, see Durham University's BASIS website at www.basisonline.org.uk.

*Breastfeeding*

The most important source of ongoing skin-to-skin contact is breastfeeding. In fact, skin-to-skin contact in one form or another is part of every single breastfeeding session. Breastfeeding is in fact a mixture of suckling and closeness between mother and baby, both of which contribute to the positive effects of breastfeeding, which will be described later.

*Babywearing*

Carrying the baby front-to-front in any kind of babywearing sling or carrier may also induce similar effects. It is not as intense as skin-to-skin, but it may be performed for a long time. For much more detail on babywearing, see *Why Babywearing Matters* by Rosie Knowles, also in this series.

### Summary

- Skin-to-skin contact is a clinical practice in which the naked newborn is placed on the mother's chest immediately after birth.
- The mother's chest skin temperature increases in response to oxytocin-induced dilation of blood vessels in the chest.
- The warmth from the mother's chest increases oxytocin release in the baby.
- Increased levels of oxytocin in the baby increase the newborn's skin temperature. The skin temperature of

mother and baby synchronises.

- The newborn performs an inborn breast-seeking behaviour involving crawling, touch, licking, breast massage and suckling.
- This interaction is boosted by maternal chest warmth caused by oxytocin.
- In the mother oxytocin levels rise in response to the baby's breast massage.
- The mother's social interaction (tactile, vocal) with the newborn is increased.
- Skin-to-skin contact decreases stress levels (cortisol levels and pulse-rate) in the newborn, and cortisol levels in mothers.
- Skin-to-skin contact stimulates functions related to milk production and growth and development in mother and infant.
- Skin-to-skin contact may induce wellbeing in mother and infant.
- The effects of skin-to-skin contact are triggered by eye contact, vocal and tactile interaction.
- Several types of sensory nerves are involved, including thin unmyelinated C- and Ct-fibres and thick myelinated fibres.
- Other situations involving close contact, such as repeated skin-to-skin later on after birth, kangaroo care, co-sleeping and even breastfeeding give rise to similar effects as skin-to-skin contact after birth.

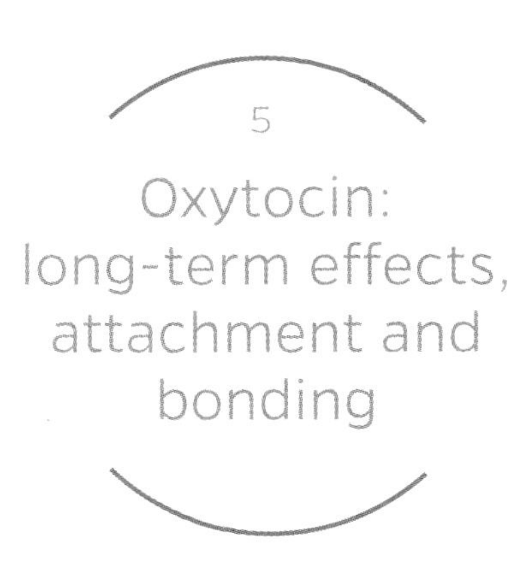


## Skin-to-skin immediately after birth is linked to long-term effects

Skin-to-skin contact between mother and baby immediately after birth is, as described in the previous chapter, related to a number of positive adaptations in mother and infant, such as enhanced social interaction and stress reduction in both of them. There is, however, an even more interesting and important aspect of early skin-to-skin contact. The effects that are induced at this point in time may become sustained. Both the behavioural effects and the physiological effects, for example the anti-stress effects, may influence both mother and baby for many years, if not for life. This will positively impact the ongoing relationship between mother and baby, as well as their future relationships in general. Also the activity of the stress system is influenced in a positive way, as well as processes related to healing.

### *Mothers who are separated from their babies after birth are less interactive and protective*

The American paediatrician Professor Marshall Klaus noted that when he investigated babies after their stay on the maternity ward, their mothers behaved differently depending on how the babies had been treated. Some babies, in particular the ones that had been born prematurely, were taken to a nursery or special care unit and had sometimes even been placed in an incubator after birth. Others had remained with their mothers on the labour and maternity wards. When he investigated the babies before they left the hospital, he noted that mothers who had been separated from their babies were less interested in what he was doing with their babies. They sat on a chair, looked out of the window and hoped that he would be finished soon. Mothers who had stayed with their babies on the ward behaved quite differently. They stood close to him and watched every movement he made.

Why should these mothers behave so differently? Marshall Klaus guessed that it had something to do with the amount of time the mothers and babies had spent together in close contact after birth. He designed an experiment in which some mothers were allowed more time in skin-to-skin contact with their babies soon after birth than others. It turned out that the group of mothers and babies who had spent more time close to each other after birth seemed to like each other more, even after several months. They looked into each other's eyes, smiled at each other and communicated vocally significantly more than the others did. These findings, which were very much doubted and criticised when they were first published, have since been replicated in many studies and are now accepted.

## The early sensitive period

Since sustained or long-term effects on social interaction

between mothers and their babies were induced when the babies and mothers had skin-to-skin contact with each other in the first hours after birth, Marshall Klaus and John Kennel, who had performed many experiments in the field, labelled this period 'the early sensitive period'.

*Long-term decrease in stress levels*

It is not only the increased social interaction between mother and baby caused by skin-to-skin contact after birth that can become sustained. Russian paediatrician Ksenia Bystrova showed that not only is social competence increased for a long time after skin-to-skin contact immediately after birth, but the baby's stress levels also decreased. She showed that skin-to-skin contact for two hours after birth increased the quality of social interaction between mother and infant and the baby's ability to handle stressful situations after one year.

*Long-term effects can also be induced later on*

It is possible to induce long-term effects by practising skin-to-skin contact later on, but the effect per time unit is weaker and has to be compensated for by a longer duration of skin-to-skin contact. There seems to be an exponential curve to describe the ability of skin-to-skin contact to induce long-term effects: the effect is strongest immediately after birth and then declines rapidly. Nonetheless, the ability of skin-to-skin contact to induce long-term effects is always there, and probably never falls to zero.

Canadian psychologist Anne Bigelow has performed studies in which skin-to-skin was carried out repeatedly with full-term babies for a few weeks after birth. She studied the effects of skin-to-skin contact on different psychological and physiological variables over a time span of nine years in both mother and baby. Expressions of increased social competence and reduced stress levels have been documented in mothers and babies, some of them even after nine years.

*How do the long-term effects of skin-to-skin contact work?*
These effects are complicated and involve many functions from sensory cues to changes in molecular and genetic mechanisms.

It is likely that the mechanisms involve activation of sensory systems, vision, sound and smell and in particular tactile stimulation, as described in the previous chapter, and consequent release of oxytocin in both mothers and babies. A very important mechanism by which oxytocin is released during skin-to-skin contact is by stimulation of sensory nerves in the skin that are activated by touch, warm temperature, pressure of different intensity and stroking. When the nerves are activated signals travel to the brain, where they release oxytocin, which in turn stimulates social interaction, decreases stress levels and stimulates growth. These assumptions are supported by a vast number of studies performed in animals.

*Extra-sensory stimulation in the newborn period gives rise to long-term effects*
Several types of study demonstrate that extra-sensory stimulation given to newborn rats may influence them for life. Extra-sensory stimulation of rat pups by soft brushing has been shown to increase the rate of growth in these animals for a long time. Blood pressure and cortisol levels may also be reduced for life by this treatment.

In another experimental model, rat pups that received more licking from their mothers (which is a type of sensory stimulation of the skin) were shown to be more social, less anxious and less stressed as adults than those that received less licking. The rats also had evidence of more robust function in the oxytocin system, as they had more oxytocin receptors in the amygdala (an important centre for control of fear and social interaction in the old part of the brain). The anti-stress effects may also be indirectly linked to oxytocin; see below.

## Role of oxytocin

### *Repeated administration of oxytocin gives rise to long-term effects*

If oxytocin is given repeatedly to female, male, young or old rats, long-term effects are induced. These effects seem to be induced by secondary effects induced by oxytocin, via activation of other signalling systems in the brain.

One important mechanism behind the long-term effects, in particular those related to stress reduction, is a decrease of activity in one of the main stress-regulating systems in the brain, the noradrenergic system. This effect is induced by oxytocin, which activates a type of receptor called an alfa-2adrenoceptor, which slows down the function of noradrenergic system in the brain and as a consequence stress levels and reactions to stress are reduced.

### *Extra oxytocin in the newborn period gives rise to long-term effects*

Similar effects of oxytocin have been demonstrated in rat pups. If newborn rat pups receive extra oxytocin for a few days, they are changed for life. They have lower stress levels, including lower blood pressure and cortisol levels as adults, and they also react less to pain. In addition, they gain weight more quickly. It is as if the effects of oxytocin become imprinted. Interestingly, blood pressure and high cortisol levels in pups that had been exposed to stress in the womb were reversed if extra oxytocin was given after birth.

*Is there a link between the long-term effects caused by extra tactile stimulation or administration of oxytocin in the newborn period?*

The long-term effects of extra oxytocin administration and extra licking or touching in adult rats, as well as in rat pups, are strikingly similar. It is therefore likely that oxytocin is in some way involved in the life-long effects that are induced by sensory stimulation of newborn pups.

From this perspective it is interesting that oxytocin acts together with or via other signalling mechanisms in the brain when long-term effects are induced. The long-term effects may therefore not necessarily reside within the oxytocin system itself, but later on in the chain of signalling events involved in the effects. Perhaps an important effect is exerted by changes in the noradrenergic system through oxytocin's effect to lower its function by activation of the alfa-2adrenoceptors as described above.

*Epigenetic mechanisms*

How is it that the effects of a short period of skin-to-skin contact extend beyond the period of newborn life? It has been suggested by Michel Meaeny and members of his research group in Canada that experiences during newborn life influence the activity of certain genes in both mother and baby. This phenomenon, in which experiences or the environment turn off or turn on the function of existing genes, has become known as epigenetics. It is likely that only very basic functions will induce epigenetic changes, such as stress levels, availability of food and perhaps of oxygen, environmental temperature and the presence or absence of supporting individuals. The existence of the long-term effects is, however, not in doubt, regardless of which mechanisms are causing them.

## Attachment and bonding

Attachment and bonding are terms that describe how babies (through attachment) and mothers (through bonding) become tied to each other after birth. For example, a newborn lamb immediately follows its mother because it learns to recognise her, and this is of course vital for survival. On the other hand, a mother also learns to recognise her own offspring. It is of course necessary that she gives milk, care and protection to her own baby. Animal experiments show that oxytocin is involved in the mother's bonding to the baby and also in the baby's bonding or attachment to the mother.

### *Mechanism of pair bonding in voles*

The best description of the mechanisms that are involved in bonding or attachment has been obtained from observations of another type of bonding, which occurs between females and males in a type of small rodent called a vole. Prairie voles normally form lifelong relationships and they stay together their whole lives. Oxytocin released in response to sex is the glue by which these voles are tied together for life.

It is, however, possible to cause bonds between a female and a male vole in an artificial way, simply by administering oxytocin into the brain of a female or male vole, in the presence of a vole of the opposite sex. If a female vole has received oxytocin in this way, she or he will prefer that vole to other voles, as shown by the American researchers Sue Carter and Tom Insel. The experience of the vole of the other sex (in the case of voles probably mostly mediated via the smell of the other individual) will be strong and easily memorised.

### *Link to reward*

The American researchers Tom Insel and Larry Young have also demonstrated that the administered oxytocin also activates the reward centre and that dopamine is released. Since this occurs at the same time as she or he experiences the odour of the individual of the other sex, that individual will be linked to pleasurable experiences not only during the first meeting, but also in the future.

*Link to reduction in stress levels*

In addition, due to oxytocin, stress levels are reduced and linked to the individual of the other sex, so she or he not only feels a sense of wellbeing when close to the other individual, but she or he will also feel relaxed. These feelings will be connected to the other individual and whenever they meet, dopamine will be released and stress levels reduced and of course she or he will be attracted to the source of pleasure and relaxation. This also means that when the two individuals are separated from each other these positive feelings vanish. What is the solution? To get back to the source of pleasure and calm as soon as possible. In other words, the two individuals have become bonded or attached to each other.

*Bonding and attachment is the opposite of aversion*

The process of bonding is in a way the opposite of aversion. If a meeting with another individual is pleasant a bond may be formed, but if the meeting is unpleasant aversion is created. Experiments in rats performed by the German researchers Inga Neumann and Werner Landgraf show that a rat which receives an electric shock when close to another rat will associate that rat with the pain and fear of the electric shock, and consequently avoid him or her in the future. The most promising part of this research is that this aversion can be extinguished, or taken away, if the fearful rat receives oxytocin in some areas of the brain in connection with meeting the rat that is linked to fear. In this way a negative reaction can be turned into a neutral or positive one.

*Imprinting*

This model of bonding is not very different from the concept of imprinting, which was coined by the Austrian researcher and Nobel Prize laureate Konrad Lorenz. He showed that a recently hatched gosling, when it comes out of the eggshell, will follow the mother wherever she goes. This is of course life-saving, because the gosling needs food and protection from its mother. However, Lorenz showed that the gosling didn't necessarily follow the mother goose, but any individual that happened to be present when it hatched. He named this phenomenon 'imprinting'. In the case of geese and other birds, vision seems to be more important than smell in the process of imprinting or bonding. Lately it has been shown that levels of the stress hormone cortisol fall when a gosling sees and follows its mother, making the processes of bonding and imprinting even more similar.

*Bonding and attachment in human mothers and babies*

Nobody has been able to measure the biochemistry of the bonding process in human mothers and babies during skin-to-skin after birth, as such research would be unethical and risk harming the individuals involved. Even if humans are in some ways more 'developed' than other mammals, the bonding/imprinting/attachment processes are, from an evolutionary point of view, very old and important for survival. There is no reason to believe that there will be any major differences in the mechanisms involved in the formation of a bond in humans compared to other mammals.

The basic elements of 'bonding' are there. Oxytocin is released in both mother and infant in response to skin-to-skin contact. Skin-to-skin contact is also associated with a decrease in stress levels, as evidenced by lower cortisol levels in both mother and newborn and a decrease in heart rate in the baby.

It is also linked to feelings of happiness and reward. Oxytocin has the potential to activate the dopamine system, which has been shown in breastfeeding mothers. These basic biological processes may facilitate the more complicated process of bonding and attachment between mother and baby.

## Secure attachment

There is another interesting consequence of skin-to-skin induced facilitation of the attachment and bonding process and the long-term physiological effects of early skin-to-skin contact, and that is how these processes are linked to the development of 'secure' attachment.

The British psychologist John Bowlby created the concept of attachment and described how this bond between children and their mothers was also linked to feelings of safety and stress reduction. In his books on attachment he also described the consequences of separation of the child and the mother. Psychologist Mary Ainsworth developed this concept further as she found that the relationship between children and their parents was different, and described the type of relationship a child had with its mother as 'secure' or 'insecure'. She diagnosed the different attachment types by studying the behaviour of children playing with their mothers and by the results of interviews and questionnaires in adults. The securely attached children were calmer and could tolerate a small period of separation from their mothers in a more optimal way than the others.

### *Type of upbringing is linked to attachment type*

It became clear that these types of attachment reflect different experiences during a child's upbringing. Individuals who are characterised as securely attached tend to have had a childhood with a lot of closeness and caring, whereas those who are insecurely attached tend to have had a less optimal upbringing.

It seems likely that the development of secure attachment starts immediately after birth. In ideal conditions, when a baby is in skin-to-skin contact with the mother, oxytocin is released in response to stimulation of nerves in the skin by touch, warmth, pressure and stroking. When oxytocin is released the entire spectrum of oxytocin effects is induced. Babies become more socially interactive, they feel well, they do not experience any fear or pain, stress levels are low and babies are relaxed.

*A link to Pavlovian conditioning*

Are you familiar with the work of the great Russian scientist Ivan Pavlov? He showed that secretion of saliva or gastric acid normally induced by the process of eating, with all its sensory components, could be linked to, or conditioned to, light or sound. He showed that if dogs were exposed to light or sound when they were eating, after a while light or sound could trigger secretion of saliva and gastric acid, even when there was no food available, after a few training sessions.

Now let's consider the newborn baby lying in skin-to-skin contact with the mother, experiencing all the positive things described above. Sensory nerves on the skin of the chest are being activated, which results in a release of oxytocin and consequently the experience of wellbeing and decreased stress levels. However, at the same time the baby is exposed to the sight, smell and sound of the mother. It is quite possible that the release of oxytocin and the consequent oxytocin-related effects triggered in the newborn during skin-to-skin contact can become conditioned. So if the skin-to-skin process is repeated many times, perhaps the baby's oxytocin release is soon triggered by only seeing, smelling or hearing the voice of the mother? And perhaps later on, the mother (or father) doesn't even need to be there, and the sensations of the mother (or father) become a constant memory in the brain, with calming and stress-reducing effects.

In some studies, adults with secure attachment have been shown to have higher oxytocin levels than those with insecure attachment. They also suffer less pain and inflammation, less anxiety and depression and are healthier in other ways. This is of course not an effect of one session of skin-to-skin contact, but of many repeated experiences of closeness. However, having skin-to-skin contact after birth may increase the likelihood of future beneficial periods of closeness, which may result in secure attachment as the baby grows up. The consequences of secure attachment for an individual's future relationships and health will be discussed in more detail in the last chapter of the book.

*The path of life is determined after birth*

Perhaps the experience of the newborn after birth to a certain extent decides their future path in life. If the external world is experienced as hostile or stressful, the individual had better prepare for that and enhance the activity of the stress system. On the other hand, if the surroundings immediately after birth are perceived as warm, supportive and pleasant, as for example during skin-to-skin contact and other kinds of closeness early in life, the pattern of effects regulated by oxytocin may be promoted. In this way calm and connection, including an increased facility for social interaction, decreased stress levels and increased capacity for growth will be prioritised.

### Summary

- Skin-to-skin immediately after birth is linked to long-term facilitation of social interaction and decreased stress levels.
- These long-term effects are most easily induced during the first hours after birth, known as 'the early sensitive period'.

- Long-term effects on social interaction and stress levels can be induced in animal experiments by extra tactile stimulation.
- Long-term effects on social interaction and stress levels can be induced in animal experiments by extra administration of oxytocin.
- The establishment of attachment and bonding between mother and newborn is facilitated after birth.
- The establishment of attachment and bonding between mother and newborn is facilitated by oxytocin.
- In animal models of pair bonding, oxytocin facilitates a link between recognition, memory, activation of reward/dopamine release and reduction of stress levels and another individual.
- The process of bonding is related to the phenomena of imprinting and conditioned reflexes.
- Skin-to-skin contact may facilitate development of secure attachment.

# 6
# Oxytocin and breastfeeding

## Breastfeeding and survival

Breastfeeding was originally the only way of feeding newborn babies. In the absence of mother's milk, babies would not survive. Today it is possible to feed babies with formula milk, but breastfeeding still is the best way to feed a newborn. Mother's milk is perfectly adapted to the baby's needs, but there is another effect of breastfeeding which is also very important and which is often forgotten in the debate about the advantages and disadvantages of breastfeeding. This is that breastfeeding gives rise to many physical and mental adaptations in mother and baby, which are mediated by oxytocin released within the brain. These effects are important, because they may positively influence the interaction between mother and baby and also the mental and physical health of mothers and newborns in the long term.

### *The original way of breastfeeding*

The original breastfeeding pattern differs substantially from the way women of today feed their babies. The first humans would have breastfed their babies much more often than mothers do nowadays, and this behaviour is still seen in some hunter-gatherer cultures that exist today. Breastfeeding might occur as often as every 20 minutes, day and night, and is possible because babies remain in close proximity to their mothers, often being carried with them during the day and sharing a sleep space at night. With such an intense breastfeeding pattern, including frequent night feeds, levels of the breastfeeding hormones, oxytocin and prolactin, are more or less constantly elevated. This maximises milk production.

*Breastfeeding as contraception*

A 24-hour pattern of frequent breastfeeding is of importance for people living as hunter-gatherers for another important reason: it is linked to contraception. Constant high levels of the milk-producing hormone prolactin inhibit the mechanisms in the hypothalamus that regulate ovulation and the menstrual cycle in women. In the absence of ovulation, no fertilisation can occur and no babies will be conceived. The oral contraceptive pill works in a similar way. This hormonal brake on fertility continues for as long as a baby continues to breastfeed frequently, which might be for a period of several years. Only when a baby stops breastfeeding does a mother's fertility return. This long period of infertility has historically been important for humans and modern hunter-gatherer cultures, who move constantly and cannot carry more than one baby at a time. An interval of 3–4 years between babies was therefore optimal, both from the point of view of the culture, and the individual mother's health.

## From breastfeeding to formula-feeding

Until relatively recently breastfeeding was more or less vital for the survival of humankind. Unless mothers breastfed, their babies would die. Of course, sometimes mothers for one reason or another didn't produce enough milk. One way to solve such problems was to receive help from other women, who were breastfeeding at the same time. Breastfeeding women helped each other by breastfeeding babies who were not their own, a kind of sharing.

*Breastfeeding without giving birth*

Giving birth has always carried risks to the mother of becoming sick or even dying, and mothers of small babies may become very sick or die for reasons unrelated to giving birth. Nature offers a solution for this problem, because it is possible to induce milk production and to breastfeed without having been pregnant and given birth, or having done so long ago. The baby's grandmother, aunt or sister, or even an unrelated woman, could thus take over breastfeeding a baby they didn't give birth to.

How is this possible? In fact, the milk-producing machinery is always present in women's breasts. High levels of hormones, including oestrogen and progesterone, during pregnancy make it easier and quicker to activate the milk-producing machinery. However, with adequate stimulation of the breasts, eating and drinking, mental relaxation and a firm belief in your competence as a breastfeeder (often supported by culture), any woman can start to produce milk. Even men can produce some milk if they try to.

This practice is still common in some cultures in Africa, where grandmothers or sisters of mothers who are sick or have died from HIV may take over breastfeeding a child. It is also good news for women who have adopted a child and want to breastfeed. The milk-producing machinery is there and it is possible to breastfeed an adopted child, even if it takes some time and effort to start milk production.

*Substitutes for human milk are available*

When humans started to have access to cows or other milk-producing animals, they of course tried to feed their babies with milk from these animals. In some cases this was successful, although the composition and quality of animal milks differs greatly from human milk. Scientific investigation of the properties of human milk has led to the development of modern infant formulas, which are 'nutritionally complete' (meaning that they contain the right amount of fats, carbohydrates and vitamins to enable infants to grow and thrive). However, all infant formula lacks many of the characteristics of human milk, including immune factors, growth factors, a vast array of oligosaccharides that feed the infant's gut bacteria and many other 'living' components. As research continues to reveal more and more about the properties of breastmilk, formula milk will of course evolve too, although it will never be able to replicate the biologically complex nature of mothers' milk.

*Why breastfeed?*

Why should women of today be encouraged to breastfeed their babies? If the simpler solution is to go and buy 'nutritionally complete' formula milk in a shop, mix it with warm water, put it into a bottle and give it to the baby, isn't it less stressful to bottle-feed and not worry about producing milk? Isn't it easier with bottle-feeding, because other people can also feed the baby? Isn't it more fair if the father helps to feed the baby day and night?

In fact there are several reasons why women of today should consider breastfeeding their babies. One reason is that it is cheap. Buying formula is expensive and may become an economical burden for mothers that don't have a lot of money. Also breastfeeding reduces the risk of acquiring infections from either the water used to dilute the formula powder, or the formula powder itself, which is not sterile. Mothers in rich economies may have access to both money and clean water, but the risks of infection from the powder remain.

Another advantage of breastfeeding is that the composition of mother's milk is superior to that of formula milk. Human milk has been designed by millions of years of evolution to fulfil all a baby's needs, and contains many ingredients that help babies grow and develop that are not present in formula milk.

In addition, breastfeeding reduces the risk of infection because breastmilk contains agents that help newborns fight the microbes that can cause disease and strengthen the baby's immune system. The first milk produced after birth, the yellowish colostrum, may even be life-saving if given to premature babies. A few drops of this milk reduces the risk of a premature baby developing necrotising enterocolitis, a severe form of inflammation of the intestines, which can be deadly.

Finally, many women simply find it easy, practical and enjoyable to breastfeed their babies. The practice of breastfeeding may also be linked to positive effects on both the mother and baby's future physical and mental health, as we will see.

## Oxytocin and milk production

Breastfeeding actually comprises two separate components: the production of milk in the breasts, and the ejection of the milk from the breast. It has long been known that two hormones, prolactin and oxytocin, are of importance for breastfeeding; prolactin because it stimulates milk production in the breast, and oxytocin because it causes the ejection of milk. When the baby attaches to the nipple and starts to suckle, sensory nerves are activated, which send impulses to the brain via the spinal cord to stimulate the release of these two hormones into the bloodstream.

*Prolactin and milk production*

Prolactin is produced in the anterior pituitary and released into the bloodstream. When it reaches the breasts, it stimulates the milk-producing cells (lactocytes), which start to produce milk.

*Oxytocin and milk ejection*

Oxytocin is, as described previously, produced in two cell groups located in the hypothalamus, the PVN and the SON. Oxytocin is transported to the posterior pituitary, and from there it is secreted into the bloodstream. When the nervous impulses initiated by the baby's suckling reach the PVN and SON, pulses of electrical activity occur in the oxytocin-producing cells and oxytocin is released into the circulation in pulses. These pulses are very short-lasting and at the beginning of a breastfeed about five pulses occur in a 10-minute period. Each of these pulses is linked to ejection of milk.

When oxytocin in the bloodstream reaches the breast, it contracts small muscles (the myoepithelial cells), which surround the alveoli where the milk is stored. In this way milk is ejected. Oxytocin also helps relax the openings of the milk ducts in the nipple. By stimulating both the contractions of the myoepithelial cells and the opening of the milk ducts, oxytocin stimulates milk ejection. This double effect of oxytocin during milk ejection resembles the effect it exerts during birth, when oxytocin stimulates uterine contractions to move the baby out of the womb, and also opens the cervix to allow the baby to move into the birth canal.

*Oxytocin promotes prolactin release and milk production*

The effect of oxytocin in connection with milk production is not restricted to milk ejection. As was mentioned in Chapter 2, nerves that contain oxytocin reach many important sites in the brain, including the anterior pituitary and the cells where prolactin is produced. Oxytocin helps the release of prolactin, and in the absence of oxytocin there will be no prolactin release and no milk production. So oxytocin is also of importance for milk production.

*The release of oxytocin is adapted to the volume of milk*
The release of oxytocin during breastfeeding is adapted to the need for milk; the more oxytocin released in response to breastfeeding, the more milk ejected. More oxytocin is released later on during breastfeeding than at the beginning of breastfeeding, because the amount of milk a baby takes at a feed increases over time and as the baby grows. In addition, more oxytocin is released in mothers of twins than in mothers of a single baby. Oxytocin levels in connection with breastfeeding are higher in mothers who have given birth to several children in comparison to those who have only one child, and more oxytocin is released in mothers who only breastfeed in comparison to those that mix breastfeeding and bottle-feeding.

*Expressing milk and bottle-feeding*
Oxytocin is released in response to expressing milk, either by the mechanical stimulation induced by a pump, or by hand expression.

Bottle-feeding itself is not associated with any release of oxytocin in mothers, as bottle-feeding is not connected to suckling.

*Different kinds of stress reduce oxytocin release*
It has long been known that milk ejection and breastfeeding may become problematic if mothers experiences stress. Indeed, several studies show that less oxytocin is released in response to suckling if the mother is exposed to stress; either an environmental stressor such as too much noise, or too

much bright light, or an inner stressor like pain or feelings of stress and fear. More unexpectedly, more subtle stimuli, such as being in an unfamiliar environment, or the mother being focussed on an intellectual task, may also reduce breastfeeding-associated oxytocin release. These findings show that mothers should ideally be in calm and familiar surroundings when breastfeeding, and also mentally relaxed, to optimise the production and ejection of milk.

These findings regarding factors that influence oxytocin release during breastfeeding are consistent with the concept of 'biological nurturing' presented by Suzanne Colson. According to her model breastfeeding will work best if mothers are completely physically and mentally relaxed.

#### *Similarities with birth*

There is a striking similarity between the factors that inhibit and promote the release of oxytocin during birth and breastfeeding. The release of oxytocin in women giving birth is not only blocked by intense fear, pain and stress, but also by subtle environmental cues such as being in an unfamiliar environment and the presence of unknown people, because this may make mothers feel unsafe. Breastfeeding women, like women giving birth, also profit from mental and physical support from other people.

### Oxytocin and physiological adaptations

Oxytocin released during breastfeeding is not only involved in the process of providing the baby with milk, but also influences the physiology of both mother and the baby in a very positive way. The reason why oxytocin may exert all these effects is that oxytocin is not only released into the bloodstream to influence milk production and milk ejection, but is also released from nerves within the brain. When the oxytocin system is activated by breastfeeding, oxytocin is released into areas of the brain involved in regulation of digestion and metabolism, of fear, pain, stress and restoration, and also of wellbeing and social interaction. All these effects help the mother adapt physiologically as well as psychologically to motherhood.

*Effects on digestion and metabolism*

Many women experience a feeling of thirst when they start to breastfeed. This is caused by oxytocin being released in areas where thirst is regulated in the brain, and as a consequence breastfeeding mothers drink more. Appetite control is also influenced, so mothers feel hungrier during breastfeeding as they must eat a little more than usual. Breastfeeding is expensive from a caloric point of view. Exclusively breastfeeding a baby 'costs' on average 800–1,000 calories extra per day. Some of these calories come from the stores mothers lay down during pregnancy, and some from food that the mother consumes.

Food is not always available in unlimited amounts. To increase the chance of successful breastfeeding and thus the chances of survival of both mothers and infants, several metabolic adaptations occur in the mother during breastfeeding. This topic was discussed more in detail in the chapter on adaptive mechanisms in pregnancy.

Each time the mother breastfeeds, when she is feeding or transferring calories to her baby, she is in a way 'eating' herself. At the same time as oxytocin is released to stimulate the production and ejection of milk, oxytocin is also released from nerves that reach areas in the brain (the vagal nuclei) from which the function of the digestive system is controlled. Each time the mother breastfeeds, hormones are released in the gastrointestinal tract to increase its size and function and promote the release of insulin, which helps us store nutrients. At the same time hormones are secreted that allow stored nutrients to move into the breasts to serve as fuel for milk production. The more milk a mother produces and gives to her baby during breastfeeding, the more the function of her gastrointestinal tract is changed.

The balance between storing and using energy is very intricate. It varies over time during breastfeeding, and the more milk the baby ingests, the more energy must be recruited from the mother's stores. It also varies between individuals. Some mothers produce less milk and tend to conserve energy during the period of breastfeeding, while others produce large amounts of milk and lose a lot of weight. For many mothers breastfeeding is the best way of using the extra calories that were put down as fat during pregnancy. Perhaps breastfeeding could even be regarded as the final stage of female reproduction, during which the mother's body returns to its pre-pregnancy state.

*Decreased activity in the stress system*

Activity in the mother's stress system, both the HPA axis and the sympathetic nervous system, is decreased during suckling via the activation of oxytocin nerves in the brain. Each time a mother breastfeeds the levels of the stress hormone cortisol decrease, because oxytocin inhibits the production of the substances that regulate the release of cortisol, both in the hypothalamus (CRF) and in the anterior pituitary (ACTH), as described in Chapter 2.

The activity of the sympathetic nervous system is also decreased by suckling, and each time the mother breastfeeds her blood pressure decreases. This is caused by inhibition of the sympathetic nervous system caused by oxytocin released in the brain.

Lower levels of cortisol and lower blood pressure reflect lowered stress levels during breastfeeding. In addition, breastfeeding mothers are buffered against stress reactions for many hours after breastfeeding. Some mothers even have trouble recruiting sufficient energy when doing physical exercise due to this inhibition of the stress system.

More importantly, some of the anti-stress effects of breastfeeding may persist long after the end of breastfeeding. They may even protect the mother against certain types of stress-related diseases later in life. This will be discussed in the last chapter of this book.

## Oxytocin and mental adaptations

There are several oxytocin-linked changes to the mother's mood, personality and ways of reacting during breastfeeding, which are linked to the increased exposure of the brain to oxytocin. These changes or adaptations are induced during each breastfeed, and some of them develop into more sustained effects that may last during the entire breastfeeding period or even longer.

### *Increased wellbeing*

When mothers are breastfeeding the reward centres in their brains are activated. Studies in which breastfeeding mothers look at pictures of their babies show that areas in the brain linked to wellbeing or reward are activated. In addition, levels of dopamine increase, more so if the baby looks happy in the picture. The same type of activation likely occurs with each breastfeed, but it is difficult to measure brain function during breastfeeding.

The rise in dopamine activity in the brain is parallel to the increase of oxytocin levels in the blood. The greater the increase in dopamine levels in the brain, the higher the levels of oxytocin in the bloodstream, which of course supports the connection between oxytocin and dopamine. Furthermore, the rise of dopamine and of oxytocin is higher in mothers with a secure attachment than in those with an insecure attachment.

*Reduced levels of anxiety and increased calm during breastfeeding*

Most mothers (but not all) report that they feel less anxious and that they are more interested in interaction with their baby and other loved ones than they usually are while breastfeeding. In addition, there is a more long-term shift in their feelings of anxiety and their interest in social interaction. A few days after start of breastfeeding mothers report that their levels of anxiety are lower and their interest in social interaction is higher than women of the same age who are not pregnant or breastfeeding.

These changes are likely to be mediated by oxytocin released in response to suckling in areas of the brain that are involved in the regulation of fear and social competence, such as the amygdala. The higher the mothers' oxytocin levels during breastfeeding, the less anxious they feel and the more interactive they are. Interestingly, the women themselves are not aware of these changes since they occur at an unconscious level.

The tendency to open up for social interaction and listen to others is to a certain extent activated during pregnancy and birth, but it becomes even more pronounced during breastfeeding once the baby is the focus of the mother's attention.

*To please and conform*

Another personality trait that may become more pronounced during breastfeeding is the wish not only to listen to others, but also to please and conform. From a biological perspective this personality trait is a mechanism by which mothers are helped 'from the inside' to care for and help other people, which includes responding to the wishes and need of others, including their own babies. There is a shift of the personality from being egoistic and self-centred to being altruistic and willing to listen to and help others.

However, there is a potential problem with this change of personality, because it can extend to people other than your baby and loved ones, for example staff in the hospital or somebody trying to sell something to you. Be cautious about signing important documents regarding money while breastfeeding!

These shifts in personality also seem to be linked to the release of oxytocin during breastfeeding. The assumption that it is linked to oxytocin is supported by the fact that mothers who are given an oxytocin drip during labour score higher in their wish to please and conform than those who are not given a drip.

*Tolerance of boring tasks*

Another helpful psychological change that occurs in breastfeeding mothers is that they become more tolerant of monotonous tasks; they don't need or wish to be in a continuously changing and stimulating environment to the same extent as non-breastfeeding women. This change may help mothers to breastfeed and change nappies many times every day.

Reduced levels of anxiety and increased levels of social interaction only occur when a mother feels that she and her baby are safe. Otherwise the mother may become more fearful and aggressive than under normal circumstances. This increased tendency to fear and aggression is described below.

## Oxytocin and maternal aggression

Oxytocin does not only trigger caring behaviours in the breastfeeding mother, but also stimulates another aspect of maternal behaviour: maternal aggression, or expressions of maternal defence and protection of the baby.

*Protective behaviours*

Maternal aggression, which is really an increased tendency to react to aspects of the environment that could potentially be dangerous for the child, has many different faces. The most clear-cut expression of maternal aggression is of course the strong physical reaction that a mother can experience when she feels that her baby is in some way threatened. This is the 'tiger mother' in action.

*Increased caution and worry*

There are also more subtle changes to the breastfeeding mother's psyche that belong to this category of reactions, which are instinctual reactions to protect the baby from potential environmental risks. Many breastfeeding mothers find it difficult and very unpleasant to experience violence of any kind. They may avoid listening to news or watching TV programmes that feature severe accidents, crime, war and other types of violence.

Some breastfeeding mothers also avoid risky situations, such as mountains, bridges and balconies where you might fall, or places with many unknown people, like markets or elevators. All these situations trigger reaction patterns that are expressions of an increased awareness of potentially dangerous situations in the environment. Some mothers may also become a little bit suspicious. They don't necessarily trust other people to hold their baby until they know that they will do so safely.

*Separation anxiety*

Many mothers also find it very difficult to leave their children alone, or to leave them at all. They avoid leaving the children because if they do it feels as if they have lost something very important during the period of separation and it causes great fear and anxiety. These feelings are expressions of the anxiety caused by being separated from the child that you are bonded to, because being with the individual you are bonded to is calming and makes you feel good, and leaving them causes anxiety and negative emotions. Leaving the child may also be experienced as risky, because something might happen to them that you can't control.

*Disturbed sleep and increased wakefulness*

It is also well known that many mothers with newborn babies have a different sleeping patterns than they did before they had their baby. Mothers wake up more easily – as soon as the baby cries or even moves – if they are sleeping near each other. Fathers also wake up more easily if they are very much involved in the care of the baby. It is important from a biological point of view that mothers or fathers wake up easily to feed and protect their babies, or to help them with something they need.

*Increased activity in the amygdala*

Why is it that breastfeeding mothers or fathers who are involved in the care of their babies become extra anxious and fearful in some situations? Why do mothers wake up so easily? From a neurophysiological point of view it is due to an increased basal activity in the amygdala, an important centre for the regulation of fear and stress reactions. During pregnancy and birth, and then during breastfeeding, the threshold for activation of the amygdala in response to some types of triggers is lowered in mothers, and it stays so for quite a long time. This is the way nature or biology creates a slight over-reactivity in mothers and sometimes fathers of newborn babies.

The best way to minimise these heightened levels of anxiety that mothers perceive is to be close to the baby, to watch it and to hold it, for example in skin-to-skin contact. The release of oxytocin triggered in these situations of closeness decreases the activity in those areas of the amygdala that are responsible for the increased wakefulness, fear and suspicion.

*Increased calm and increased fear, two sides of the same coin*

High oxytocin levels in connection with birth and also in response to skin-to-skin contact and breastfeeding are responsible for both the calming effects and the increased anxiety and wakefulness. They are just two sides of the same coin. As stated before oxytocin has a double effect profile. The amygdala is a bit over-reactive and there is an increase in the basic protective function of the amygdala during parenthood. That's why small things in the environment that normally doesn't cause anxiety suddenly do. Activity in the amygdala is then blocked when mothers and fathers are close to their baby and when the surroundings are calm. Put another way, both the exaggerated anxiety and the exaggerated calm experienced by parents are expressions of being attached to or loving someone.

Exactly when the over-reactivity of the amygdala subsides is not known, but it probably remains for quite a while and parents have to live with it.

*Temporary shift between emotional and intellectual intelligence*

Some mothers also experience a shift in their mental capacity or intelligence, especially during breastfeeding. It is obvious that they become more emotional. Their 'emotional intelligence' (EQ) is increased, and with that comes an enhanced ability to read facial expressions, interpret tone of voice and in other ways tune in not only to the baby, but also to people in general.

At the same time some women find that their more traditional, logical intelligence is somewhat compromised. It may be difficult to express linear, logical thoughts for a while, and it is well known that it may be more difficult to learn and retain the memory of non-interesting things like numbers and letters. It just doesn't seem important. These effects are all caused by the flood of oxytocin in the brain that occurs every time a mother breastfeeds. The question is, how long does this last?

## Problems during breastfeeding

Breastfeeding rates vary between countries and breastfeeding isn't always easy. In principle there is no link between the biological capacity to breastfeed and problems with breastfeeding. Low breastfeeding rates and breastfeeding problems have different roots.

However, this is not a book about breastfeeding practices, but about oxytocin. Since oxytocin plays an important role in breastfeeding, it is important to find practices and recommendations which allow maximal oxytocin release. As stated above, oxytocin release is promoted by environmental and inner calmness and relaxation. Information about how to breastfeed, and that breastfeeding is natural and normal from people you trust is important, but most of all it is important to trust in your own capacity to breastfeed. Again safety and trust are linked to oxytocin. 'Biological nurturing', as described and taught by Suzanne Colson, is in line with these ideas.

### *Depression and anxiety*

The association between depression/anxiety and breastfeeding is complex. As Kathleen Kendall-Tackett has shown, if mothers are depressed after birth, which is not uncommon, breastfeeding might be compromised. In fact, depression influences oxytocin release in the same way as stress: oxytocin levels in response to breastfeeding may be lower. On the other hand, breastfeeding, especially when it is linked to skin-to-skin contact, may also in some situations counteract depression, as it is linked to a decrease in stress levels. Oxytocin release during suckling and skin-to-skin contact is linked to a powerful decrease in levels of cortisol, a decrease in blood pressure and other signs of relaxation. These effects tend to become stronger with time. So if depressed mothers continue to breastfeed, paradoxically they may be helped out of their depression and oxytocin levels may rise as stress levels decrease.

*Dysphoric milk ejection reflex*

D-MER, or dysphoric milk ejection reflex, is an uncommon and unpleasant reaction that some women experience while breastfeeding. Milk ejection is perceived as unpleasant and linked to negative feelings. Why this sometimes happens is not known. Some theories have linked the effect to a dip in dopamine levels in connection with prolactin release. This is not likely to be the case, however, due to the timing of symptoms. Symptoms occur within minutes after the start of breastfeeding, in connection with milk ejection. This suggests that the symptoms are somehow linked to oxytocin. Normally oxytocin induces calm, wellbeing and relaxation, but it seems as if these effects are not being expressed in women with D-MER. Rather, it suggests that there is a mistake in the wiring of the oxytocin-linked effects in the brain normally associated with milk ejection. Further, a very powerful decrease in blood pressure and blood sugar levels caused by oxytocin in connection with milk ejection might be experienced as unpleasant and contribute to the negative experience.

## Oxytocin in connection with expressing milk or bottle-feeding

*Expressing milk*

In some countries it is quite common for breastfeeding women to use a breast pump or hand expression to extract their milk. Then they give their own milk by bottle to their babies. This is a common practice if women have to return to work soon after birth and continue to breastfeed.

It is sometimes more difficult to get the milk out with a breast pump than via breastfeeding. Some mothers have found that if they look at a picture of their baby when they are expressing it works better. It is important to stimulate milk production and ejection with all of your senses!

The suckling stimulus provided by expressing induces some of the effects caused by the baby when breastfeeding. Oxytocin is released into the blood, as is the milk-producing hormone prolactin. Stress levels are also reduced by expressing, as the levels of the stress hormone cortisol and ACTH, the hormone that stimulates cortisol secretion, decrease. It is likely that some other adaptive oxytocin-linked effects are induced as well, but research is still lacking in this area.

*Milk ejection can be developed into a conditioned reflex*

Normally, milk ejection is triggered when the baby suckles. After a while milk ejection develops into a conditioned reflex and is triggered when the mother sees, hears or thinks of her baby, or sometimes another baby. That is what is happening when a mother's milk lets down unexpectedly when she is shopping in the supermarket without her baby, but hears someone else's baby cry. Sometimes mothers who feed their babies by expressing milk may become 'attached' to the breast pump, and find it more difficult to breastfeed the baby directly, because the pump rather than the baby triggers their milk ejection reflex.

*Bottle-feeding*

In mothers who bottle-feed there is no oxytocin release in connection with feeding. This means that bottle-feeding mothers are not exposed to all the oxytocin effects induced in the brain, including anti-stress effects and the mental adaptations to motherhood, when they are feeding their babies. Some of these effects can be obtained if bottle-feeding mothers hold their babies in close skin-to-skin contact when feeding.

There are several studies showing that there are some differences in the way breastfeeding and bottle-feeding mothers interact with their babies. Bottle-feeding mothers have been shown to be less 'sensitive' to their babies when they interact with them, and they also have higher levels of sympathetic nervous tone and lower levels of parasympathetic tone. These differences are due to the lack of repeated exposure to oxytocin that occurs in the brains of breastfeeding mothers.

### Oxytocin and the baby during breastfeeding

What about babies? Are they influenced by oxytocin during breastfeeding? Yes, in many ways. They are indirectly influenced by oxytocin released in the mother during breastfeeding. The amount of milk and warmth mothers transfer to their newborns is in part a consequence of oxytocin release in the mother during suckling and skin-to-skin contact during breastfeeding. Also, the mother's caring behaviour and the sensitivity of her social interaction is promoted by maternal oxytocin.

#### *Production of oxytocin in the breasts*

A mother transfers some oxytocin to her baby via breastmilk, because oxytocin is produced in the breasts and secreted into the milk. So each time a baby breastfeeds they will receive some oxytocin. These levels are, however, relatively low, and as oxytocin is not well absorbed from the gastrointestinal tract it will not contribute in a significant way to the baby's blood levels of oxytocin. This milk-derived oxytocin might, however, have positive effects on the mucosal lining in the gastrointestinal tract.

#### *Suckling and oxytocin release in calves*

It is easier to observe the release of oxytocin in response to suckling in other species. Oxytocin levels rise when a calf is suckling. If the calf drinks the same amount of its mother's milk from a bucket, oxytocin levels don't rise. These results show that it is not the amount of ingested calories, or something in the milk, which activates oxytocin release in the drinking calf, but the suckling action. How is the release of oxytocin during suckling brought about? Sensory nerves in the mucosa of the mouth of the calves are activated when they are suckling on the cow's udder.

*Stimulation of digestion and growth*

As a consequence of the suckling stimulus, oxytocin, released in the brain, stimulates the vagal nerve and the levels of hormones involved in digestion and metabolic processes, including the release of the energy-storing hormone insulin, are increased. In this way more efficient storage of nutrients occurs and calves gain more weight per ingested calorie when they suckle, because the whole network of hormones and nerves in the gut is activated.

*Suckling and human babies*

As we have seen, babies produce their own oxytocin, both before they are born and afterwards. Some oxytocin is released into the circulation in response to the baby suckling. It has also been shown that ingested energy is used more efficiently after breastfeeding than after bottle-feeding. These and other data show that the baby's suckling stimulates the vagal nerves and exerts positive effects on digestion, metabolism and growth. In addition, pain relieving and sedative effects have been demonstrated in response to breastfeeding.

*Sucking on a pacifier induces calm and stimulates weight gain*

Some studies show that sucking on a pacifier, or dummy, induces similar effects. One effect of sucking a pacifier, which is very easy to observe, is the calming effect. Most parents have noticed that they can calm a crying baby almost instantaneously by giving them a pacifier unless their babies are too hungry or too angry.

The energy-saving effects caused by oxytocin and the activation of the vagal system can also be induced by sucking a pacifier. Studies show that sucking a pacifier stimulates weight gain. A baby who receives food through a tube (bolus-feeding) gains more weight if they are allowed to suck a pacifier at the same time as being bolus-fed, than if they are not sucking a pacifier. This is because the nerves and hormones of the gut are activated in response to the sucking stimulus.

This effect pattern is related to the oxytocin-linked effects caused by soft and pleasant sensory stimulation of the skin. Babies who had skin-to-skin contact also gained more weight when bolus-fed than those babies that did not.

## Summary

- Oxytocin is released in pulses in response to the suckling stimulus of breastfeeding.
- Oxytocin in the bloodstream induces milk ejection.
- Oxytocin facilitates the release of prolactin and thereby milk production.
- Suckling-induced oxytocin release in the mother's brain decreases stress levels by decreasing the activity in the HPA axis and the sympathetic nervous system.
- Suckling-induced oxytocin release in the mother's brain increases the activity of digestive and metabolic functions by increasing the function of the vagal nerve.
- Suckling-induced oxytocin release in the mother's brain may promote wellbeing by activation of reward mechanisms.
- Suckling-induced oxytocin release in the mother's brain makes mothers less anxious.
- Suckling-induced oxytocin release in the mother's brain

makes mothers more socially interactive.

- Suckling-induced oxytocin release in the mother's brain makes mothers more pleasing.
- Expressing milk is also linked to oxytocin release and stress-relieving effects.
- Bottle-feeding is not linked to oxytocin release.
- Levels of anxiety may be increased during breastfeeding due to increased activity in the amygdala.
- Suckling gives rise to oxytocin release in the newborn and to sedative, pain-relieving and metabolic effects.

# 7 Oxytocin and medical interventions during birth

Today spontaneous or physiological vaginal birth, i.e. vaginal birth without medical interventions, is becoming less common in many countries, as the use of different types of medical interventions during birth such as caesarean section, epidural analgesia and intravenous infusion of oxytocin increases.

Caesarean section was originally used as an acute treatment to save life during complicated deliveries, but nowadays both acute and planned caesarean sections are used more and more often and for a wider spectrum of indications.

For reasons which are only partially known, many women of today are afraid of giving birth and in particular of the pain of giving birth. For this and many other reasons different types of medical interventions have been developed. It is even a right in law in many countries that women should have access to pain relief during labour and birth.

If a woman is having a vaginal birth, epidural analgesia is often given to decrease the experience of pain during labour. Epidural analgesia often results in a need for infusions of oxytocin to augment contractions during labour, and in addition oxytocin infusions are often given to initiate labour before spontaneous labour has actually started. Today epidural analgesia is by far the most common type of pain relief in connection with birth.

The aim of this book is not to judge or to give recommendations as to which type of birth a woman should have. It is a book about the importance of oxytocin in connection with birth. As described in detail in the chapters above, the release of oxytocin during and after birth, in connection with labour and in particular during skin-to-skin contact, induces positive, short- and long-lasting positive adaptations in mother and baby. Oxytocin plays an important role in all these adaptations. It stimulates all types of positive interactions between mothers and infants, and helps them to bond with each other. In addition, it reduces their stress levels and improves their health. From this perspective it is therefore important to know whether medical interventions in any way influence the oxytocin release normally induced in connection with spontaneous labour and birth. If oxytocin release is reduced as a consequence of medical interventions, the mother and her baby may not experience the positive effects that oxytocin induces during normal birth, so what are the consequences? Is there any way to compensate for any losses? We will discuss these questions in this chapter.

## Caesarean section

Caesareans were originally performed to save the lives of mothers and babies in situations when labour was not progressing normally. Today this type of caesarean section, known as an emergency caesarean section, is still performed when there is an immediate need to bring the baby out quickly, but the indications have widened.

There is also another type of caesarean section, known as an elective caesarean section. In this case the caesarean is performed before spontaneous labour has started. These caesarean sections may be performed because the baby is too big to be born vaginally, or because of the position of the baby in the womb. This type of caesarean is becoming more and more common. Sometimes a mother-to-be does not want to have a vaginal birth, perhaps because of fear of the pain of birth, or of the long-term consequences of birth (for example changes to the body, the shape of the vagina, future incontinence). In some cases mothers may have a strong desire or need to give birth at a particular time (for example to start treatment for a medical condition).

The use of elective caesareans may be promoted by labour wards and hospitals because it is considered more efficient to know when women will give birth. In some countries practical and economic reasons may increase the rate of elective caesareans. Performing caesareans may also increase the speed at which women give birth if the capacity of a labour ward is insufficient.

The frequency of caesarean sections is on the increase and in some places almost 90% of women give birth in this way. But what happens when so many women give birth by caesarean? Obviously not all of these caesareans are carried out for life-saving purposes.

In these circumstances the question must be asked whether there are negative consequences for the mother and baby of elective caesarean, because when they are performed in such numbers we are looking at a much larger group of women, who do not need a caesarean to save their life. What is good in an emergency situation may not necessarily be as positive under more normal circumstances: the balance between positive and negative consequences may change.

*Differences in oxytocin between emergency and planned caesareans*

In terms of oxytocin there is a fundamental difference between emergency and elective caesareans. Mothers who have an emergency caesarean section may have had an almost normal labour and birth, with a caesarean performed at the end. In this situation oxytocin may be released in amounts that are almost the same as during a spontaneous vaginal birth.

In contrast, a planned or elective caesarean is performed before spontaneous labour occurs. Therefore, there will be no oxytocin release.

*Both mother and baby have lower oxytocin levels after caesarean*

If oxytocin levels are measured immediately after a caesarean, both mothers and newborns have lower oxytocin levels than mothers and babies after a spontaneous vaginal birth. This difference is particularly clear when mothers having a normal spontaneous birth are compared with those having an elective caesarean. These data support the idea that both the mother and the baby are exposed to less oxytocin in connection with caesarean than in spontaneous vaginal birth, and consequently experience fewer effects of oxytocin.

*Delay in skin-to-skin contact*

Some other factors differ between normal physiological labour and both types of caesarean. In emergency and elective caesareans mothers receive spinal or epidural anesthesia, which may affect oxytocin release in a negative way after birth in connection with skin-to-skin contact between mother and baby (which will be described below). In addition, mothers who have an emergency caesarean are at risk of a delay in having skin-to-skin contact after birth in comparison to mothers who have a vaginal birth. Because of this delay they may not have their babies close to them during the early sensitive period. Awareness of the potential consequences of this delay has been growing, and it is becoming more common for skin-to-skin to be facilitated in the operating theatre. Many mothers choose to make their preference for early skin-to-skin part of their plans for an elective or emergency caesarean.

*Fewer oxytocin pulses in response to breastfeeding*
Are there any negative consequences of caesarean which may be associated with a decreased release of oxytocin? Yes: it seems that oxytocin release in response to breastfeeding is affected. Normally, a mother releases up to five peaks of oxytocin in the blood in connection with breastfeeding two days after birth. In contrast, mothers who have had a caesarean only release up to one peak, and this difference is associated with decreased secretion of the milk-producing hormone prolactin. The amount of milk the baby receives is also reduced. These effects may make breastfeeding and bonding with the baby more difficult for mothers and babies who have had a caesarean, unless they receive extra support and help with breastfeeding.

*Delayed development of maternal adaptations*
As we have seen, oxytocin is not only released into the circulation during birth, but is also released into the brain. The oxytocin released during labour and in connection with birth influences the mother's psychology in such a way that her social skills improve and her levels of anxiety decrease. These changes do not occur two days after birth in mothers who have had a caesarean, simply because of the lack of, or reduced release of, oxytocin during the birth.

*Reduced release of oxytocin in response to skin-to-skin contact*
Another perhaps surprising finding is that mothers who have had an elective caesarean do not release oxytocin in response to skin-to-skin contact and suckling immediately after birth. How is this possible? The most likely explanation is that oxytocin released during birth makes the skin extra-sensitive to contact with the baby. The increased sensitivity of the skin is one of many preparations for interaction and bonding between mother and baby, which oxytocin released during birth and skin-to-skin contact normally induces.

This means that skin-to-skin contact after birth may not be as beneficial for mothers who give birth by elective caesarean as it is for mothers who have had a vaginal birth, unless the mothers have received an infusion of oxytocin to stop bleeding from the womb after birth. This will be described more in detail below.

## Pain relief

The pain of labour is not a new problem and all cultures have developed ways to help women cope while giving birth. Reducing pain during labour is not only about decreasing the subjective sensation of pain in mothers giving birth; it may also help the progress of labour and birth, since extreme pain and fear may inhibit the progress of labour. Pain and fear can inhibit oxytocin release and increase function in the sympathetic nervous system.

*Traditional knowledge based on closeness and support*
The most natural and physiological way of inhibiting pain during labour is to increase the release of oxytocin. Traditionally, this has been accomplished by closeness and support provided by other women. These women historically didn't know anything about oxytocin and its stimulatory effect

on uterine contractions, its effect on social interaction and its anti-stress effects. They acted intuitively based on experience and traditional knowledge. This old way of inhibiting pain and speeding up the process of labour is now subject to a revival, and will be discussed later on as an alternative to pharmacological pain relief.

In the first half of the 20th century women were often anaesthetised when their baby was born to save them from the pain. This technique is not used any more. Later on injections of drugs with pain-relieving effects, such as different types of opioids (for example, morphine, pethidine or meperidine) were given and they sometimes still are.

Local anaesthetics were also introduced. These drugs, such as marcain, were applied locally around nerves that transmit pain from the uterus and vagina during birth. Paracervical and pudendal blocks are examples of this type of treatment.

## Epidural analgesia

Today epidural analgesia, or EDA, is by far the most common pain-relieving technique used by birthing women. Epidural analgesia consists of an opioid, such as fentanyl or sufentanil, and a local anaesthetic such as marcain. This mixture is injected into the epidural space outside the cover (the dura) of the spinal cord in order to block the activity of nerves mediating pain before they enter the spinal cord. The ways in which epidurals are administered, as well as their composition, have developed over the years and there are many different types of epidural. Epidural analgesia very efficiently reduces pain in childbirth and its use is increasing rapidly all over the world.

### *Reduced levels of oxytocin*

It is, of course, of great importance to reduce excessive pain during birth, as intense pain and fear may stop the progress of labour. In this way epidurals may facilitate the progress of labour. Epidural analgesia may, however, also have a counterproductive effect on the progress of birth, as in addition to blocking the nerve fibres from the uterus and the vagina that mediate pain, it may also block the nerve fibres that give rise to oxytocin release (the Fergusson reflex). Mothers who have received epidural analgesia have lower levels of oxytocin in their circulation, because less oxytocin is released than during an unmedicated birth.

*Reduction in uterine contractions*

It is well known that uterine contractions may become weaker and less frequent in women who have received epidural analgesia. This reduction in uterine contractions is likely to be caused by the reduced release of oxytocin in women who have had epidural analgesia. Reduced oxytocin release leads to a reduction in uterine contractions and a slowing down of labour.

This may in turn increase the need for other medical interventions, such as forceps or a caesarean. Often an infusion of synthetic oxytocin is given to the woman giving birth in order to increase uterine contractions, which were not optimally stimulated in the absence of sufficient amounts of endogenous oxytocin.

However, this is not the only problem with epidural analgesia. The release of oxytocin in the brain is also reduced by epidural analgesia. Oxytocin in the brain stimulates behavioural and physiological adaptations during birth, as described in more detail in Chapter 4, and these effects do not occur after administration of epidural analgesia.

The release of the milk-producing hormone prolactin is decreased in response to breastfeeding in the first days after birth, which may explain why women who have received an epidural sometimes have problems with breastfeeding.

*No oxytocin-mediated pain relief during labour*

One consequence of epidural analgesia is that the physiological pain-relieving effect of endogenous oxytocin during labour is partially blocked, as the release of oxytocin is reduced. In fact, the normal built-in pain-relieving effects of the body's own oxytocin during birth are replaced by the pain-relieving effect of the epidural itself, which blocks the transmission in the pain fibres. In addition, opioids are one of the strongest inhibitors of oxytocin release in the brain, and therefore the opioid content of epidural analgesia may also reduce the secretion of oxytocin in this way.

*No amnesic effect*

The lack of oxytocin in labouring women who have received epidural analgesia may explain yet another interesting finding. If women are asked, some time later, about the intensity of pain they experienced during birth, those women who had a normal vaginal birth report the pain of birth as less intense than immediately after labour, and more importantly less intense than those mothers who gave birth with the help of epidural analgesia. How is this possible? The most probable explanation is that oxytocin, in addition to having pain-relieving effects, also has an amnesic effect, so mothers who had a normal birth without epidural analgesia experienced the pain-relieving effects of their own oxytocin as well as the long-term amnesic effect that makes mothers forget the pain of birth.

*Less pain, but not increased wellbeing*

After birth, women who have had a vaginal birth often report an intense feeling of joy and happiness. This is not always the case in women who have received epidural analgesia. They do not report such high levels of satisfaction and wellbeing after birth as women who have not had epidural analgesia.

This difference in the perception of wellbeing occurs in spite of the clear reduction in pain experienced by these mothers, suggesting that wellbeing is not a consequence of pain relief. Instead wellbeing could be a consequence of oxytocin released into the brain, with activation of the reward centre and dopamine release. This would also imply that the mothers who have had epidural analgesia don't experience the consequences of the oxytocin-induced activation of the reward centre and dopamine release.

Nurses and midwives who work in labour and maternity wards report that mothers who have had epidural analgesia are sometimes less engaged in their birth experience. Birth becomes more like an everyday event and not a peak experience. Women don't seem to need to talk about their birth experience as much as women who have had a vaginal birth. It may take a little longer for these women to take in what has happened and adapt to motherhood.

*Less transmission of warmth from the chest*

When babies are in skin-to-skin contact with their mothers and breastfeeding, their skin temperature increases. This is a response to the oxytocin-mediated transmission of warmth from the mother's chest. In breastfeeding babies whose mothers received epidural analgesia two days earlier this temperature response is lacking. These results indicate that epidural analgesia may block the oxytocin-mediated increase in maternal skin temperature. This reduced transmission of warmth from the maternal chest may also explain why newborn babies perform less hand massage during skin-to-skin contact with mothers immediately after birth if mothers have received an epidural. It seems as if the lack of transmission of warmth from mother to baby during skin-to-skin contact after birth is somehow remembered by the baby two days later.

*Delay in maternal psychological adaptations*

As we have already discussed, the oxytocin released during birth starts the process of changing the mother's brain to become adapted to motherhood. For example, mothers become less anxious and more socially interactive. The development of these adaptations is delayed in mothers who receive epidural analgesia during birth, as well as in mothers who have a caesarean. This means that the oxytocin-related maternal adaptations develop later if a mother has had epidural analgesia or given birth by caesarean section than in mothers who have had a vaginal delivery without epidural analgesia.

Again there seems to be an antidote to this delay in the development of maternal adaptations. Mothers who receive an infusion of synthetic oxytocin during labour do better than those without an infusion: their psychological adaptations resemble those induced after a vaginal delivery without epidural analgesia. In addition, this temporary delay of maternal psychological adaptations is overcome later on in breastfeeding mothers who are exposed to frequent oxytocin release.

## Summary

Caesarean section:

- The use of caesarean, both emergency and planned, is increasing.
- Less oxytocin is released into the circulation in both mothers and babies during caesarean, in particular during elective caesarean, which is not associated with uterine contractions, than during vaginal delivery.
- Less oxytocin is released in response to skin-to-skin contact immediately after birth in mothers who had an elective caesarean.
- This effect can be restored by infusions of synthetic oxytocin postpartum.
- Less oxytocin is released a few days after birth during

breastfeeding in mothers who have had a caesarean.
- Less prolactin is released a few days after birth during breastfeeding in mothers who have had a caesarean section.
- These effects may in part explain why caesarean section may be linked to difficulties with breastfeeding.
- Oxytocin release within the brain is also reduced following caesarean. This may explain why mothers who have had a caesarean do not develop psychological adaptations with decreased anxiety and increased social interactive skills like mothers who have had a normal vaginal birth.
- This effect can be restored by infusions of synthetic oxytocin postpartum and also, after some time, by breastfeeding.

Epidural analgesia:
- The use of epidural analgesia to decrease pain and anxiety during birth is increasing and epidural analgesia is a powerful inhibitor of pain during birth, which may be advantageous for the progress of birth.
- Epidural analgesia acts by blocking the nerve fibres mediating pain from the uterus and vagina within the epidural space. Epidural analgesia may also block the nerve fibres mediating oxytocin release (the Fergusson reflex). This reduces oxytocin release, and diminished oxytocin release leads to a reduction in uterine contractions and a slowing down of labour, which may lead to other medical interventions.
- Oxytocin release within the brain is likely to be reduced by epidural analgesia. This may explain why mothers who have given birth with epidural analgesia lack the amnesia for pain experienced by mothers who have had a normal vaginal birth without epidural.
- It may also explain why mothers who have given birth with epidural analgesia report less wellbeing and satisfaction after birth than mothers who have had a normal vaginal birth without epidural.
- Mothers who have given birth with epidural analgesia do

not develop psychological adaptations of decreased anxiety and increased social interactive skills.

- Newborns exhibit less interaction during skin-to-skin contact if their mothers have received an epidural, or just the local anaesthetic marcain in significant amounts, perhaps as a consequence of decreased maternal emission of warmth.
- Babies born to mothers who had epidural analgesia two days earlier do not experience an increase in skin temperature during breastfeeding, indicating a lack of maternal signalling to the baby even a few days later.
- Infusions of synthetic oxytocin may counteract some of the negative psychological effects caused by epidural analgesia alone. However, the combination of oxytocin and epidural is not always positive.

# 8 Synthetic oxytocin

Oxytocin was one of the very first human hormones to be described from a chemical point of view. Oxytocin has been produced (synthesised) and used clinically during labour since the 1960s. Synthetic oxytocin is sometimes called exogenous oxytocin, Pitocin or Syntocinon.

Infusions of synthetic oxytocin are used to initiate and augment uterine contractions during labour and birth. In addition, injections or intravenous infusions of oxytocin are often given immediately after birth to induce contractions of the uterus in order to prevent bleeding after birth. The use of oxytocin for these indications is increasing all over the world and in some countries 90% of all mothers receive exogenous or synthetic oxytocin during birth.

## Administration of synthetic oxytocin

More and more often oxytocin is given to augment labour if it doesn't proceed as expected.

One common reason for the dramatic increase in the use of synthetic oxytocin in connection with birth is the increased use of epidural analgesia. As described in the previous chapter, a side-effect of epidural analgesia is that it decreases the release of endogenous oxytocin and consequently the intensity of uterine contractions.

There are also other reasons for the increased use of intravenous infusions of oxytocin during labour. There is an increased demand from both mothers and healthcare professionals to initiate or induce labour and birth at a specific time point before its natural onset, and nowadays birth is initiated by intravenous administration of oxytocin, in some labour wards as a routine.

Very often an injection of oxytocin is given routinely to reduce bleeding after birth.

*Debate regarding clinical effects*

There is an ongoing debate about the effectiveness of the use of synthetic oxytocin in connection with birth. The effects of synthetic oxytocin may be less than previously assumed, and may depend on other circumstances around birth. The duration of birth may be shortened by a few hours by administration of exogenous oxytocin, but the use of forceps or caesareans is not reduced. The clinical recommendations for the use of synthetic oxytocin are not, however, within the scope of this book; they remain the responsibility of obstetricians and midwives to resolve.

*Negative side-effects?*

What is of interest in this book is the question of possible side-effects following administration of synthetic oxytocin. Many women find that infusions of oxytocin give rise to more painful contractions. It has also been suggested that infusions of oxytocin can increase the frequency of anxiety and depression in the mother after birth and damage brain function in the unborn baby. As will be discussed below, these effects, if they exist at all (most of the studies suggesting these negative consequences are not of high quality), are likely to be indirectly caused by the unphysiological pattern of uterine contractions caused by infusions of synthetic oxytocin.

*Oxytocin levels during natural birth and following administration of exogenous/synthetic oxytocin*
Even though the oxytocin that is produced in the hypothalamus and released into the circulation during labour and birth and the exogenous/synthetic oxytocin given as an intravenous infusion are one and the same substance from a chemical point of view, their mechanism of action may differ in several ways.

*Plasma levels of oxytocin*
During normal vaginal birth oxytocin is released in pulses of increasing frequency and size. A maximum frequency of three pulses per 10 minutes is reached towards the end of labour. In connection with birth a giant peak of oxytocin is induced as a response to the Fergusson reflex. In contrast the plasma profile of oxytocin after an intravenous infusion of synthetic oxytocin is flat and sustained.

*Amounts of oxytocin infused*
Normally oxytocin infusions during labour are first administered at a low rate, which is gradually increased until effective contractions are observed. At the beginning of the infusion blood levels of oxytocin are not much higher than those observed during normal physiological labour, but as the rate of infusion is increased oxytocin levels in the blood might become three to four times higher than is normally seen in spontaneous labour. It is quite easy to calculate the increase in oxytocin concentration in the blood in response to increased doses. If the infusion rate is doubled, the plasma level doubles, and so on.

Protocols for administration of oxytocin are not well standardised and differ substantially between countries. The amount of oxytocin and the speed at which it is administered might therefore be higher than necessary in some countries. This is important, since if intravenous infusions of oxytocin during labour are associated with negative side-effects, they are likely to be related to the dose given and possible negative effects may be avoided by decreasing the dose.

When oxytocin injections or infusions are used to stop bleeding from the uterus after birth, large amounts of oxytocin are given in a short time: often 5 or 10 IU of oxytocin is given as an injection after birth. The same amount of oxytocin may be given during eight hours of infusion of oxytocin during labour. This explains why the levels of oxytocin in the mother are only moderately (3–4 times) higher during infusions of oxytocin during labour, but of course much higher for a short while after injection of the same amount after birth.

*Oxytocin levels and the pattern of uterine contractions*

The oxytocin pulses observed during normal delivery are in part responsible for the contractions of the uterus, which help the mother to give birth to the baby. The flat oxytocin level induced by infusions of synthetic oxytocin gives rise to a different pattern of uterine contractions. It is a common observation that the contractions obtained after exogenous administration of oxytocin are not of exactly the same type as those occurring during normal birth. The contractions induced by infusions of exogenous oxytocin appear to be more long-lasting and the mother experiences them as more painful. Depending on the dose of infused oxytocin, the number of contractions, and their strength and duration, may exceed those seen during normal labour.

*Effects of oxytocin in the maternal brain during birth*

During spontaneous physiological labour oxytocin is not only released into the circulation, but also from nerves within the brain. Oxytocin is, as described above, extremely important for the process of birth, but it is also a hormone that exerts adapative effects in the brain. It has pain-relieving effects and most likely exerts anti-stress effects during birth; in the absence of oxytocin the experience of pain and stress levels would be higher. Oxytocin during birth also prepares the mother for motherhood, as it increases her social skills and lowers her levels of anxiety. Most likely the high oxytocin levels in connection with birth are linked to an activation of the reward centre and dopamine release, which add to the mother's experience of joy after birth. In addition it has an amnesic effect, making the mothers forget the pain and other negative memories of birth.

In contrast, infusion of synthetic oxytocin does not induce exactly the same effects in the brain. This is because infused synthetic oxytocin remains in the circulation, because the blood-brain barrier prevents its transfer from the circulation to the brain. Only minor amounts of oxytocin will pass the blood-brain barrier in response to oxytocin infusions within the normal dose regimen used. There are, however, other indirect ways by which infused oxytocin can influence the brain. Infusions of synthetic oxytocin may induce some of the effects normally induced by oxytocin released in the brain during birth, by enhancing the activity of the Fergusson reflex via its stimulating effects on uterine contractions.

*Consequences of strong uterine contractions*

When the muscles in the uterus are contracting, the brain is informed by increased activity in the sensory nerves mediating pain. Contractions observed in response to infusions of exogenous oxytocin are stronger, more long-lasting and consequently more painful than normal physiological contractions. The increased signalling in the sensory nerve fibres from the uterus that mediate pain results in increased activity in the stress system. This is a negative consequence of oxytocin infusions. Not only because mothers experience pain and thereby fear and anxiety, but because the effects may become long-lasting. Often epidurals have to be given to women who receive oxytocin infusions, because of painful contractions.

*Risk of long-term effects because of the early sensitive period*
As we have seen, the period during and just after birth has been labelled the early sensitive period, because effects induced during this period tend to become long-lasting. Some studies suggest an increased risk of anxiety and depression in women who receive infusions of synthetic oxytocin. These studies have to be interpreted with caution, since they often lack information on the doses of oxytocin infused and the concomitant use of epidurals. However, if such negative consequences of oxytocin infusions do exist, they may be a consequence of the increased stress levels that are induced by such infusions. In response to an aberrant pattern of uterine contractions, characterised by too long and too frequent contractions, which are also painful, the mother's stress levels will be increased during labour. These increased stress levels may turn into a more chronic state, which will later on facilitate the development of anxiety disorders and depression.

Indeed, some expressions of increased stress levels in mothers who receive intravenous infusions of synthetic oxytocin during labour have been observed in the days after birth; their levels of the stress hormone cortisol are higher than in those not receiving oxytocin.

*Possible positive effects of oxytocin infusion*
Somewhat surprisingly, infusions of oxytocin during and

after labour have been linked to a strengthening of some mental changes that normally occur after vaginal delivery and breastfeeding. As mentioned above, oxytocin released in the brain during birth may change the maternal personality to favour social interaction and reduce anxiety. Interestingly, some of these traits, related to increased sociability, seem to be affected by infusions of exogenous oxytocin. Mothers who receive infusions of synthetic oxytocin report more pleasing social behaviours and less anxiety than mothers who received less oxytocin some days after birth. Infusions may even restore some of the negative effects on these maternal adaptations induced by epidural analgesia. This is likely a consequence of the increased activity in the Fergusson reflex and consequent increase in mother's own oxytocin release that is induced following enhanced uterine activity.

*Infusions of oxytocin restore effects lacking after caesarean*

An even more surprising finding is that the absence of these mental adaptations seen in mothers who have given birth by planned caesarean are restored by infusions of synthetic oxytocin after birth. Not only are these mental adaptations restored, but also the oxytocin release caused by skin-to-skin contact and suckling immediately post-partum, which are not expressed in mothers who have had a planned caesarean. These data could be interpreted as showing that oxytocin is needed during birth to increase the sensitivity of the skin. This may be one way in which social contact between mother and infant, and bonding between them, is facilitated after birth. If there was no release of oxytocin in connection with birth, as is the case in a planned caesarean, the effects can be restored by an infusion of oxytocin after birth given in order to prevent postpartum bleeding.

*No transfer of oxytocin to the baby*

It has been suggested that infusions of synthetic oxytocin into the bloodstream of the mother, given in order to initiate or augment labour, will pass through the placenta and enter the baby's bloodstream. Some suggest that this type of secondary oxytocin infusion might influence the function of the baby's brain in a negative way.

This chain of events is not likely to occur for several reasons, the most important being that the chemistry of oxytocin makes it difficult for oxytocin to pass from the placenta into the baby's circulation, and then later on to pass from the baby's circulation into the baby's brain. The oxytocin levels in the mothers receiving oxytocin are higher than during normal conditions, but not high enough to allow any significant transfer of oxytocin from the mother to the infant through the placenta. Furthermore, the unborn baby has as high or even higher oxytocin levels than the mother, making it even more unlikely that the infused oxytocin would travel from the mother to the baby.

*Indirect actions*

However, even if the oxytocin infused into the mother's circulation doesn't pass directly into the baby, there are other ways in which oxytocin infusions could influence the unborn child. The baby may be indirectly and negatively affected if the uterine contractions are too long and strong. In such situations uterine blood flow may be reduced, and transfer of oxygen to the baby may be compromised. Also, very strong and frequent uterine contractions may be experienced as painful for the baby in the womb.

*Oxytocin, a dangerous drug?*

Oxytocin is listed as one of the most dangerous drugs in the world. The reason for this is that if oxytocin is strongly overdosed during labour, uterine contractions will become too strong and frequent and blood flow to the placenta may be reduced, meaning that the baby will suffer from insufficient oxygen. This of course is a very rare event, and only occurs in response to very high doses of oxytocin given for a long time.

*The baby produces its own oxytocin*

As mentioned above, the baby has its own oxytocin, and the release increases during labour and birth. It helps the baby to endure the stress of being born. Oxytocin most likely reduces pain in the baby and it also, together with a release of adrenaline, minimises inflammation and prevents slight tissue damage normally occuring during birth. However, it is possible that the release of oxytocin in the baby follows the same rules as the release of oxytocin in the mother: it may be inhibited by very intense stress. When the baby's own oxytocin levels are lowered the baby's brain will then be more susceptible to damage by hypoxia or too low glucose levels.

## Controversy regarding the use of oxytocin

There is a big debate about whether synthetic oxytocin should be used to induce or augment labour during birth as often as it is, and also about the increase in its use. There may be situations when oxytocin is needed to promote labour or reduce bleeding, but some studies indicate that the effects of oxytocin infusions are limited to a relatively minor shortening of labour, and do not decrease the frequency of caesarean sections and other medical interventions. It is therefore possible that too much oxytocin is given too often. Of course the more often a drug is used, the more likely it is that potential side-effects will come to the surface. The cost/benefit ratio will definitely be changed.

## How to compensate for the lack of oxytocin release after caesarean and epidural?

*Advantages to medical interventions*

Sometimes it is necessary to have a caesarean, even if the mother wanted to have a normal vaginal delivery. Complications do occur during birth, which require an emergency caesarean. In these cases we should be extremely grateful that the surgical techniques for caesarean have developed so much, and that caesarean carries minimal risks for mother and baby.

Likewise, sometimes there are good reasons for performing an elective caesarean. The mothers may be very afraid of giving birth, the baby may be too big or lying in an awkward position, or the baby may be breech. In some clinics doctors and midwives prefer to carry out a caesarean when the baby is breech, because the skills to deliver breech babies vaginally are declining.

For some women epidural analgesia is a great relief because it gets rid of unbearable pain. It may also promote the progress of labour when the mother's pain and fear have stalled it.

*Association with reduced oxytocin levels*

However, there are also caesareans and epidurals that are not so necessary, but for different reasons are performed anyway. These are linked to a reduced amount of oxytocin during birth. As we saw above, emergency caesareans may be associated with oxytocin released while labour progressed normally. In this way emergency caesareans can resemble normal birth. Elective caesareans are different because they are not associated with onset of labour and oxytocin release. The administration of epidural analgesia reduces the release of oxytocin during birth, since not only the nerves that mediate pain, but also the fibres mediating oxytocin release, may be blocked.

*Role of stress levels*

Mothers who have had an emergency caesarean may have reduced oxytocin levels for another reason. These mothers may have high stress levels, since they are often exposed to intense stress before the caesarean was performed. As we know, high stress levels lead to reduced oxytocin levels. Remember that there is an antagonism between the stress system and the oxytocin system in the hypothalamus. When stress levels are high, oxytocin release is reduced, and when activity in the oxytocin system is high, the function of the stress system is reduced.

Mothers having a planned caesarean may not be so stressed before the caesarean is performed, but irrespective of the type of caesarean all mothers are subjected to surgery and consequently have a wound in the uterus and in the skin. This means that having a caesarean, like any type of surgical intervention, is linked to activation of the stress system. When the skin or any other organ is exposed to trauma, the sensory nerves that mediate stress reactions and pain are activated from the traumatised area.

From this point of view, epidural analgesia is different. Epidural analgesia blocks the release of oxytocin, but also inhibits stress responses by efficiently reducing pain and fear. In a sense it may reduce stress levels a bit too much.

*Consequences of low oxytocin levels*

The reduced oxytocin levels caused by emergency and planned caesareans and epidural analgesia give rise to a certain oxytocin 'deficiency' in mothers after birth. They release less oxytocin during skin-to-skin contact, possibly as a consequence of decreased sensitivity in the skin. As oxytocin is linked to increased blood flow and increased temperature in the skin of the mother's chest, these interventions are likely to reduce the transmission of warmth from the mother's chest to the baby. As a consequence, newborns receive less warmth during skin-to-skin contact after birth, which slows their approach to and interaction with the mother, and their own increase in temperature. In other words, the mother's warmth boosts the activity of the newborn who is in skin-to-skin contact with the mother, an effect which is less prominent after planned caesareans and epidural analgesia.

Mothers may also experience less joy and satisfaction after birth, and their psychological adaptations to increased social interaction and decreased levels of anxiety are not so well developed.

Breastfeeding is also linked to a reduced release of oxytocin and/or prolactin during breastfeeding, in mothers having had caesareans or epidural analgesia, which of course may lead to problems with initiation of breastfeeding. In a way the treatment of this oxytocin deficiency is relatively easy. Oxytocin release simply has to be stimulated in every possible way.

*Extra skin-to-skin contact*

From this point of view, the routine of skin-to-skin contact becomes very important in mothers who have given birth by caesarean section or who have had epidural analgesia. The more skin-to-skin the better, and not only immediately after birth. Some studies suggest that the mother's skin may be quite insensitive during the first hour after birth due to lack of oxytocin and perhaps as a consequence of the spinal or epidural analgesia they have received. So a prolonged or even slightly delayed period of skin-to-skin contact should be used. Any extra closeness with the newborn baby will be beneficial.

*Extra support and touch*

The midwife, doula, partner or other close friends should provide mothers with as much support, warmth and touch as possible. The more mothers feel seen and supported in this way, the more their fear will disappear and feelings of safety and trust in themselves and others will be increased. This way of increasing oxytocin levels may be of extra importance in this situation, because it is likely that the links to oxytocin release from the skin are more compromised than the mental ones, due to the surgical procedures and anaesthesia the mothers have received.

*Breastfeeding*

Breastfeeding is one of the most oxytocin-rich periods in life. Oxytocin is released in mothers, but also in babies in response to each breastfeed. Therefore, breastfeeding can be used to overcome the consequences of too little oxytocin being released during labour, birth and skin-to-skin contact. The repeated oxytocin pulses can heal the damage of loss of oxytocin.

As the release of oxytocin in response to breastfeeding is compromised after caesarean, these mothers, like mothers who have had an epidural analgesia, may have problems initiating breastfeeding. But with some extra support and help these initial breastfeeding problems may be overcome. To breastfeed is not only to give milk; it is also to give warmth and touch, which also helps to compensate for the lack of oxytocin during birth.

*Baby massage*

Many mothers who have had a caesarean or epidural analgesia, who want to stimulate or enhance their interaction and bonding with their baby, may profit from learning baby massage. Massage boosts the oxytocin system in both mother and baby.

*Infusions of synthetic oxytocin*

Surprisingly, infusions of synthetic oxytocin may help restore some of the oxytocin-linked effects lost after caesarean and epidural analgesia. Oxytocin infusions given during labour help the maternal psychological adaptations to develop; mothers become more socially interactive and calmer. Also infusions of oxytocin given immediately after birth influence the mother in a positive way. The lack of oxytocin release in response to skin-to-skin treatment in mothers who have had a planned caesarean is restored, and these mothers also develop the mental adaptive changes which are otherwise lacking after caesarean section and epidural analgesia.

Newborns whose mothers have received infusions of synthetic oxytocin approach their mothers more during skin-to-skin contact, and their skin temperature increases more than in those mothers who did not receive extra oxytocin. These findings support the important role of oxytocin in transmission of warmth to newborns, and its positive consequences for the newborn's physiology and behaviour.

### *Negative effects of oxytocin infusions*

Oxytocin infusions are not currently used to induce the positive effects on mothers and babies discussed above. The results that show these positive effects are obtained from studies in which the mothers had received infusions of synthetic oxytocin for other purposes, for example to augment labour or in order to inhibit bleeding after birth.

Also it seems that the combination of epidurals and infusions of synthetic oxytocin during birth is linked to severe inhibition of oxytocin release, an effect that can't be predicted by the effects of the interventions alone. The mechanism behind these combined negative effects is not yet understood.

## Summary

- Infusions of synthetic oxytocin are used to initiate and augment uterine contractions during birth.
- Injections or infusions of oxytocin are used to stop bleeding from the uterus after birth.
- Normal labour is associated with a pulsatile release of oxytocin.

- Infusions of synthetic oxytocin result in flat oxytocin levels. This gives rise to a different pattern of uterine contractions; these may become more long-lasting and painful.
- The levels of oxytocin in response to infusions of oxytocin may be 3–4 times higher than the levels during normal physiological labour.
- Injections of oxytocin given postpartum give rise to much higher levels of oxytocin.
- Oxytocin infusion is often given together with epidural analgesia.
- Oxytocin is released from nerves into the brain during normal physiological labour.
- Infused synthetic oxytocin does not pass the blood-brain barrier. However, infused oxytocin may influence brain function by increasing the activity in the Fergusson reflex.
- It is not likely that oxytocin infused to the mother during labour can pass into the baby's circulation in any significant amounts.
- The long-lasting and painful contractions of the uterus induced by oxytocin infusions may influence mother and baby.
- Infusions of oxytocin may exert some positive effects: they may enhance maternal psychological adaptations, oxytocin release in response to skin-to-skin contact and maternal chest skin temperature.
- The combination of epidural and oxytocin infusion gives rise to effects that can't be explained by either intervention alone.
- Mothers who have given birth by caesarean section or with epidural analgesia may have a certain lack of oxytocin.
- Oxytocin release should be stimulated in every possible way in these mothers.

# 9
# Why oxytocin matters

As we saw at the beginning of this book, oxytocin was originally thought to be a maternal hormone. Later on it was labelled the hormone of love, but the aim of this book is to show that it should really be called the 'hormone of health and life'.

When mothers are pregnant and give birth to and breastfeed their babies, oxytocin release and function is increased and the effects of oxytocin are more clearly expressed than in other situations in life. This makes it a valuable period in which to study oxytocin and its effects, not only because of the important effects in women in the short and long term, but because maternity can be regarded as a means to study the function and effects of an oxytocin system that occurs in humans of all ages and both sexes, albeit to a lesser degree.

## Oxytocin during female reproduction, a model for oxytocin effects

*Oxytocin is continuously released during pregnancy, labour, birth, skin-to-skin and breastfeeding*

Oxytocin is released and exerts important actions during

all phases of female reproduction, during pregnancy, labour and birth, during skin-to-skin and during breastfeeding. The 'intention' of oxytocin is always to help the mother with the process of giving birth: it makes it possible for her to allow the baby to grow inside her during pregnancy, to give birth and create life, to bond with the baby during skin-to-skin contact after birth and to nourish it during breastfeeding.

Some of the oxytocin effects are specific for single phases of female reproduction: for example, oxytocin induces contractions of muscles in the uterus to transfer the baby from the womb to the outer world during labour, and contracts the muscles in the alveoli of the breasts to induce milk ejection and transfer of milk to the baby during breastfeeding.

Other oxytocin-related effects are of more a general character and occur, although in slightly different ways, during all the phases of creating a child. One of the main effects of oxytocin in the brain is to adapt the mother's behaviour and emotions to parenthood. Her competence at social interaction is enhanced; she becomes more open to interaction with other people, her capacity to understand nonverbal communication develops and she becomes more sensitive to others' needs and a little bit more willing to listen to and please other people. During pregnancy and birth this increased wish and need for social interaction is directed towards other adults and after birth the mother's own baby is of course the focus of her attention.

The anti-stress effects of oxytocin are activated in the beginning of pregnancy, during skin-to-skin contact and during breastfeeding; the activity in her stress system (the HPA axis and the sympathetic nervous system) decreases, resulting in lower blood pressure. These adaptations help the mothers save energy and also calm them. For obvious reasons the stress system is activated during labour.

Digestive and metabolic effects with a focus on saving and storing energy are induced by an oxytocin-mediated increase in vagal nerve activity at the beginning of pregnancy, during skin-to-skin contact and during breastfeeding. Towards the end of pregnancy and during breastfeeding a more mixed pattern of metabolic effects, including both storage and recruitment of energy, are activated. In addition, other restorative functions are activated, including potent anti-inflammatory effects.

*Breastfeeding might compensate for lack of oxytocin release during birth*

It is even possible that the oxytocin release induced in fully breastfeeding mothers might substitute for a lack of oxytocin during labour in mothers giving birth by caesarean or those having a vaginal birth and epidural analgesia. The massive amounts of oxytocin released during breastfeeding may be sufficient to induce the long-term oxytocin-linked adaptations to social interactive skills and reduction of stress and healing effects in both mother and baby.

### Skin-to-skin contact in fathers also increases activity in the oxytocin system

Are fathers and mothers alike? Not exactly. Mothers have gone through pregnancy and given birth, and have therefore already started to bond with the baby. Their oxytocin systems are preactivated by high oestrogen levels during pregnancy and by the intense release of oxytocin during birth and breastfeeding, which facilitate interaction with the baby and bonding.

Fathers do not give birth or breastfeed, but they can still have skin-to-skin contact with their babies and be close to them in many ways. They can quickly acquire the social skills needed to interact with the baby in a sensitive way. This means that their own oxytocin system can be activated and that they also stimulate the baby's oxytocin system.

The increased social competence of parents caused by oxytocin released by skin-to-skin contact is of course be of great importance for mothers and fathers and the way they interact with their babies. It will help them create a happy and secure baby.

### Long-term effects of breastfeeding in mothers

Normally blood pressure increases over the years, but women who give birth to a second child have lower blood pressure than when they gave birth to their first one! This is because the powerful anti-stress effects of oxytocin during pregnancy, skin-to-skin contact and breastfeeding become long-lasting.

Each time a woman breastfeeds oxytocin is released into her circulation and her brain. This means that during breastfeeding oxytocin effects are induced and mothers become less anxious and more socially interactive. In addition, their stress levels fall, which is expressed by a fall in the levels of the stress hormone cortisol and blood pressure, and their digestion and metabolism is influenced to optimise the use of energy for milk production.

After six weeks of breastfeeding there are signs of more sustained anti-stress effects, as the mother's blood pressure is lower at this time than at the beginning of the breastfeeding period. As mentioned above oxytocin, in particular after repeated administration, can give rise to long-term anti-stress effects, which are likely to be caused by a decreased function in the central noradrenergic stress system, an effect which is mediated by an oxytocin-induced increased function of alfa-2adrenoceptors.

Several studies have found that women who have breastfed have a different disease pattern from those who have not breastfed. What is evident from the literature is that mothers who have breastfed their babies are partially protected from a number of diseases. Long after breastfeeding, mothers who have breastfed have a 'dose dependent' (number of breastfed children times the duration of breastfeeding of each child) reduction in the risk of developing cardiovascular disease, such as high blood pressure, heart infarction, stroke and type 2 diabetes. They also have a reduced incidence of rheumatoid arthritis and breast cancer.

*Oxytocin may mediate the effects*

The reduction in the incidence of all these types of diseases may in part be secondary to the effects caused by the massive release of oxytocin that occurs during breastfeeding. It is likely, but not yet shown, that the high levels of oxytocin during birth and skin-to-skin contact after birth contribute to these positive health-promoting effects. The decreased risk of developing high blood pressure, heart infarction, stroke and type 2 diabetes are likely to be linked to the oxytocin-linked long-term reduction in stress levels, including lowering of blood pressure and perhaps also to the anti-inflammatory and restorative effects. The decreased risk of having rheumatoid arthritis may be a consequence of long-term anti-inflammatory effects and the decreased risk of some types of cancer might be linked to the capacity to normalise some types of cancer cells.

### Long-term positive consequences of closeness and breastfeeding in the newborn

As described above, both mother and baby (and possibly also the father) can be influenced for life by closeness after birth. Social interaction will be enhanced in mothers and babies and stress levels will be decreased in the baby for a very long time. These effects are induced by the surge of oxytocin that occurs during skin-to-skin contact. Obviously there are other factors that help or maximise the effects of oxytocin immediately after birth, since as little as one to two hours of skin-to-skin contact after birth suffices to induce the long-term changes. For example, the levels of the maternal hormones oestrogen and progesterone are very high, as is the activity in the central noradrenergic system, which facilitates learning. Skin-to-skin contact later on, such as during the first weeks after birth, also induces a similar effect pattern: social interactive skills are promoted and stress levels reduced, but the time needed to induce the effects is longer.

Taken together, the more closeness mothers and fathers and their babies have with each other, the more the oxytocin-linked sustained effects will be developed. Skin-to-skin contact immediately after birth, skin-to-skin contact during the first weeks of breastfeeding and breastfeeding by itself cause the same types of effects. The oxytocin released during pregnancy, labour and birth probably also contributes to these long-term effects. All these oxytocin-rich situations contribute to a long-term increase in the activity of the oxytocin system and thus an enhanced ability for social interaction, lowering of stress levels and stimulation of systems linked to growth and restoration. In this way 'virtuous cycles' can be promoted, with more social interaction and low stress levels.

*Secure attachment*

The term 'secure attachment' is a psychological term used to describe individuals who handle social interaction and stressful situations well. These individuals have often received plenty of closeness and good care from their parents or carers when they were young. Individuals with insecure attachment of different kinds have more problems with social interaction and are often more easily stressed, and they have often had a less advantageous upbringing (there might also be genetic factors). The interesting aspect from the perspective of oxytocin is that the characteristics of individuals with secure attachment are reminiscent of the effects that are induced by repeated activation of oxytocin release early in life. Such individuals have good social skills and are calm and handle stressful situations well. They seem to have a better health profile than individuals who have a less secure type of attachment; they seem to be less prone to anxiety and depression and they are less likely to develop disease with pain and stress-related disorders.

Another consequence of secure attachment or an oxytocin-linked personality profile is that the likelihood of good care of the next generation is increased. Parents with good social skills will help their children develop good social skills and also lower their stress levels and increase their capacity for restoration and healing, and thereby the children's potential for mental and physical health.

## The imbalance between stress and oxytocin-related effects in our society

Vaginal birth, as well as skin-to-skin contact and breastfeeding, may help create secure attachment between the mother and her newborn. In addition, vaginal birth, skin-to-skin contact and breastfeeding may promote mental and physical health in mothers and babies.

A beneficial consequence of interaction between parents and their babies is the development of their future health. All the oxytocin-related effects induced during skin-to-skin contact and breastfeeding (less inflammation and pain, less fear and anxiety, less activity in the stress systems, more activity in the systems leading to growth, restoration and healing) are positive for both mental and physical health.

It is important to experience periods of calm and relaxation, because this is linked to restoration of body and mind. But perhaps this is of greater importance today than ever before. The reason for this is that the environment is more stressful now than it used to be. More things happen, smart phones allow immediate access and demands for answers to telephone calls, text messages, emails and other social media. The stress system is bombarded with input and it becomes overactive. This results in an imbalance between the stress (fight or flight) system and the oxytocin (calm and connection) system in favour of the stress systems, with all the negative consequences.

Not only is there less time for relaxation nowadays, there is also less input to the calm and connection system. We spend more time alone or distanced from our surroundings, without positive social contact with others, which leads to under-stimulation of the oxytocin system. Also, interactions with computers and smartphones do not give rise to oxytocin-linked relaxation and stress reduction in the way that normal social interactions between individuals do. In this way the stress load is further increased and the capacity for social interaction is reduced. In this way a vicious circle is promoted: more stress–less social interaction–more stress etc.

*Health promotion for life*

The oxytocin system optimises social interaction, reduces stress levels and stimulates healing and growth all through life. So with a maximal stimulation of the oxytocin system comes secure attachment, better relationships, less anxiety and depression, less pain and inflammation and less stress. All of these effects will promote health.

As oxytocin induced by skin-to-skin contact and breastfeeding seems to be an important mediator of positive effects on social interaction and decreased stress levels throughout life, it is very important to promote oxytocin release and to identify which factors facilitate oxytocin release and oxytocin-induced effects in the early stages of life.

## How to maximise the effects of oxytocin during birth?

In a calm, familiar and safe environment, oxytocin is released and stimulates social interactive behaviours and gives rise to calming effects and also to stimulation of healing and growth. But if the environment is tough and experienced as threatening either oxytocin release is extinguished, or the oxytocin system will initiate a stress response.

In order to maximise oxytocin release and the positive outcomes of oxytocin mothers should try to have a normal birth. In connection with birth all kinds of stressors should be avoided and instead factors that stimulate oxytocin release should be introduced: not too much sound or light, presence of supportive people, mental relaxation, trust in yourself.

Then the mother and or the father should be in skin-to-skin contact with the baby immediately after birth. Again privacy and a calm and friendly environment is necessary.

Then the mother should breastfeed. More or less the same recommendations apply for birth, skin-to-skin and breastfeeding. No stress, the presence of supportive people, closeness to the baby and frequent suckling are important, as is trust in your own capacity.

### *The environment matters*

It is well known that stress during birth, skin-to-skin contact and breastfeeding is deleterious. Severe pain and fear, as well as environmental stressors, unsafe surroundings or the presence of a frightening person may stop the release of oxytocin and thereby inhibit the processes of giving birth or milk. The inhibition of oxytocin release may of course also result in decreased development of the positive long-term oxytocin-linked effects.

*The doula principle*

What is not so well known and accepted is that the opposite is also true. A very friendly, warm and supportive environment can actively promote these processes. An environment that is perceived as safe will help mothers release oxytocin. In addition, the presence of friendly and supportive people (often women) who help mothers relax, and feel well and safe, promote the release of oxytocin in mothers giving birth. This has often been referred to as the 'doula principle' and it involves both physical components such as touching and mental qualities such as the ability to convey feelings of empathy, warmth and support.

The 'doula principle' is in fact of more general nature. Breastfeeding mothers also thrive when they feel supported by people who really wish them well, and skin-to-skin contact after birth will not result in positive effects if mothers feel frightened or embarrassed.

The 'doula principle' probably applies to all aspects of the care system. Nobody who feels unwelcome, misunderstood or frightened will receive all the benefits the care system is supposed to give them. The 'doula principle' should therefore be extended to include all aspects of the medical care system.

The presence of friendly, empathic, warm and supportive people in connection with birth not only decreases the risk of developing negative psychological birth experiences, but may also turn a normal physiological birth into a very positive experience, perhaps one of the most memorable times in a woman's life.

*Support and warmth in connection with birth may help resolve previous traumatic experiences*

Skin-to-skin contact after birth may reduce stress levels and counteract the risk of developing anxiety, depression or even PTSD. Breastfeeding may also contribute to these healing effects. This is because of the large amounts of oxytocin released during birth and skin-to-skin contact and after birth during breastfeeding. The release of oxytocin during all these phases decreases stress levels, and the release of oxytocin is enhanced if women feel secure and safe. To be in an unfamiliar environment or to be surrounded by people who are not perceived as friendly will be linked to a reduction in oxytocin release and thus to reduced oxytocin effects.

As described in detail above, warm and friendly support during birth and skin-to-skin contact decreases the risk of developing long-term negative psychological effects after birth, because these procedures stimulate oxytocin release which reduces stress reactions. Perhaps more surprisingly it has even been shown that support and breastfeeding may heal previous traumas! It's as if repeated exposure to large amount of oxytocin, like during birth, skin-to-skin treatment and breastfeeding, can open up old memories and then, in a warm and supportive environment, 'dissolve' or extinguish the pain or stressful memories, contributing to better mental health. Studies involving the administration of substances that stimulate the release of oxytocin support the idea that oxytocin may be linked to the extinction of stressful memories.

## How to minimise the effects of oxytocin during birth

### *Absence of birth, skin-to-skin contact and breastfeeding*

Just as it is possible to maximise exposure to oxytocin, it is possible to minimise it. Not to give birth vaginally, to have medical interventions in connection with birth, not to have skin-to-skin contact after birth, as well as not breastfeeding, will lead to a reduced exposure to oxytocin in mother and baby.

Does this matter? It depends, but as we have seen, exposure to oxytocin can give rise to several types of positive effects, which may also become long-term effects that are beneficial for both parents and children. Pregnant women should be informed about these positive consequences of having a natural birth. Perhaps this knowledge would encourage normal vaginal birth and decrease the number of births taking place with medical interventions.

*Stress during birth*

Giving birth vaginally is no guarantee of high oxytocin levels during birth. If the environment is perceived as stressful, threatening or even unfamiliar, oxytocin release in the mother will be reduced. Sometimes this is the case in connection with births in hospitals. Even if awareness of the negative effects that stress might have on the birthing process is increasing, the solution is often to use medical interventions rather than focussing on reducing stress levels and supporting women, thus helping the mother's own oxytocin system to operate in an optimal way. The knowledge we have about the positive effects of oxytocin in connection with birth and skin-to-skin contact in the short and long term should be highlighted throughout medical care; currently, it is better understood by midwives than obstetricians.

*Angels or animals*

Not so long ago it was completely impossible to talk about biology and motherhood. Everybody had been taught that all feelings have a psychological or sociological background and to imagine or suggest that there were any biological roots to how mother and baby interact was like heresy. It was simply not possible that humans could be similar to animals. Partly because of religion we considered ourselves to be more like angels.

Today most people admit that humans do have some instinctual behaviours, and that we are part of an evolutionary process. There is simply no reason why we should not share some of the instincts or inborn behaviours and adaptations that other animals, in particular mammals, have.

Of course society plays an important role, and we are taught how to handle our infants by family, medical professionals and the internet. In this way our instinctual behaviours have been masked, and our inborn adaptations of a psychological and physiological nature have not always been allowed to be expressed. Modern maternity care practices that are more natural and facilitate activation of our natural behaviours and reactions are now beginning to be reinstated, although against a backdrop of pressure on health services to reduce costs wherever possible.

*Biology and guilt*

Some people refuse to accept the importance of biological adaptations in parents and newborns. They do not accept that normal birth, skin-to-skin contact between mother and baby after birth and breastfeeding can create effects of importance for the mother and baby, and the father, in the future. They do not believe that these procedures can promote secure bonding and attachment between parents and babies, which in turn may facilitate interaction between mother/father and child and affect how the child will handle his or her relationships as an adult. They refuse to see the connection between early biological processes and possible consequences for an individual's long-term health and wellbeing. Thus the slight adaptations in the mother, which are there to help mothers show their babies how to interact socially, to handle stress and to develop into a secure individual, are forgotten or denied by some people in our society. One reason seems to be to protect parents from any feelings of guilt. Those who deny any role for biology are very difficult to understand. Acknowledging our biology can help us to ensure that we protect and support mothers/fathers and babies in our society.

*Birth, skin-to-skin contact and breastfeeding and equality between men and women*

Truths and facts which are scientifically proven are sometimes denied for another reason: in the name of equal rights between women and men. Some people view equality between men and women in the labour market to be of such prime importance that any slight extra bond between mother and newborn baby is seen as a hindrance in women reaching their goals of becoming successful and achieving high positions in their professional life. For such people breastfeeding might become a problem, and for reasons of equality some mothers refrain from breastfeeding. Such people also in a way refuse to accept the complexity of being a mother and the difficulty of allowing the child to influence your life. What they definitely don't want to take in or accept is that it might matter for the future of the child. Today many fathers take a very active part in caring for the baby from birth. In relationships in which parents are 'equal' and both help with the baby, the baby's needs should still be at the centre, even if it means that the mother and father contribute by doing different things when the baby is small. If not, there is a danger that the needs of the baby are forgotten.

## Why oxytocin matters

Wouldn't it be stupid not to cultivate our oxytocin system for future generations? For this reason, it may be important to encourage vaginal birth, skin-to-skin contact and breastfeeding, as these situations are linked to oxytocin release and powerful long-term oxytocin-linked effects.

The way mothers and fathers (and of course other people involved in the care of a baby) interact with the child, in particular when it is newly born, matters because the baby's oxytocin system may be stimulated or not. If the baby's oxytocin system is stimulated, that baby will likely grow into an adult who will treat their own baby the same way. Thus a positive oxytocin cycle is created.

Good and solid function in the oxytocin system, which is established at a young age, will in most individuals be linked to secure attachment, which increases the chances of having positive relationships and stable mental health later in life. In addition it contributes to better physical health. This is why oxytocin matters.

## Summary

- The same oxytocin system optimises social interaction, reduces stress levels and stimulates healing and growth all through life.
- The release of and effects of oxytocin that occur in women in connection with labour/skin-to-skin contact/ breastfeeding could be regarded as a model for the effects of the oxytocin system.
- Skin-to-skin contact between fathers and babies also increases activity in the oxytocin system.
- Oxytocin may induce long-term effects in breastfeeding mothers with increased risk of cardiovascular disease many years later.
- The release of oxytocin in the newborn induced during birth/skin-to-skin contact may improve social interactive skills, reduce stress reactions and facilitate functions related to healing and growth in the long term.
- Love and care given to the newborn increases social skills and reduces stress levels in the adult individual. Exposure to oxytocin is linked to creation of secure attachment and in the long run to the possibility of better mental health. Exposure to oxytocin is also linked to effects that may promote physical health.
- An individual exposed to love and care will also treat his or her children in the same way. In this way an intergenerational oxytocin cycle is created, because in turn the children will do the same to their children.

# Acknowledgements

The main message in this book is to demonstrate that oxytocin is important, not only because it stimulates uterine contractions during birth and milk ejection during breastfeeding, but also because oxytocin has such a wide spectrum of associated beneficial effects.

I want to thank all the people around the world who share my interest in the unique properties of oxytocin, and who work to spread knowledge about normal female and male physiology in connection with pregnancy, birth and breastfeeding. They know that supporting natural birth and breastfeeding is not only about giving old-fashioned recommendations to women of today, but is also about bringing forth, from an evolutionary perspective, old, powerful effects that our culture has rejected or forgotten and which should be of importance in our time. I am one individual in a chain of researchers, often but not always female, who works to bring forward and upgrade our inborn, intuitive knowledge.

I want to thank all my former students and collaborators from preclinical and clinical research. I learnt so much during these processes. In particular Maria Petersson and Linda Handlin (preclinical), and Ann Marie Widström, Eva Nissen, Anna-Berit Ransjö-Arvidsson, Marianne Velandia, Ksenia Bystrova and Wibke

Jonas (clinical), but there are so many others.

Over the years I have met so many inspiring and pioneering women all over the world. I discovered some of my first soulmates in England, including Françoise Freedman, Amali Lokugamage, Clare Willocks and Suzanne Colson. Michel Odent deserves a special star in the list of important people who have inspired me. He has been and still is a pioneer within this field. As a consequence of meetings he organised I met so many other enthusiastic people, including Susan Arms, Wendy Kline, Laura Uplinger, Pascale Pagoda, Henrik Norholt, Ann Bigelow and many others from whom I have learnt so much.

I would also like to thank Soo Downe, who created and supervised the COST Action IS1405 on natural birth. Being part of these meetings has broadened my understanding and knowledge about birth and breastfeeding practices and given me a network of wonderful colleagues and friends all over Europe, and in fact all over the world. The COST Action also inspired me to initiate a number of review articles on oxytocin-linked effects in humans, work that has been invaluable for me and for this book. I would like to give special thanks to my main co-authors within the COST project: Anette Ekström and Sarah Buckley, Marie Berg, Anna Dencker, Zada Pajalic, Jean Agius Calleja, Deidre Daly, Ibone Olza, Stella Aquarone, Claudia Meier Magistretti, Sandra Morano, Claudia Massaretti, Mechtild Gross, Alicjia Kotkowska, Luise Lengler and Karolina Luigemaier.

I would also like to thank Kathleen Kendall-Tackett for our collaboration and so many inspiring high-level conversations and Kicki Hansard, Uta Strait and Yasue Ota, who helped me translate my books into English, German and Japanese.

All these people have understood that often mothers, fathers and babies might profit from having a natural birth, skin-to-skin contact after birth and breastfeeding, because this allows innate biological patterns and mechanisms, often linked to oxytocin, to be expressed, which help parents and the newborn. They know that nature has a plan, which is in the service of mothers, fathers and babies, society and life.

# References and further reading


Al-Saqi SH, Jonasson AF, Naessén T, Uvnäs Moberg K. 'Oxytocin improves cytological and histological profiles of vaginal atrophy in postmenopausal women'. *Post Reprod Health*. 2016 Mar;22(1):25-33.

Bergman NJ. 'Birth practices: Maternal-neonate separation as a source of toxic stress'. *Birth Defects Res*. 2019 Jun 3. doi: 10.1002/bdr2.1530.

Bergman NJ. 'Mother Care in African countries'. www.ncbi.nlm.nih.gov/pubmed/26303808 *Acta Paediatr.* 2015 Dec;104(12):1208-10.

Bigelow AE, Power M, MacLean K, Gillis D, Ward M, Taylor C, Berrigan L and Wang X. 'Mother-infant skin-to-skin contact and mother-child interaction 9 years later'. *Social Development* 2018; 27;937—951

Binder P, Gustafsson A, Uvnäs Moberg K, Nissen E. 'Hi-TENS combined with PCA-morphine as post caesarean pain relief.' *Midwifery*. 2011 Aug;27(4):547-52. doi: 10.1016/j.midw.2010.05.002. Epub 2010 Jul 7.

Bosch OJ, Young LJ. 'Oxytocin and social relationships: From attachment to bond disruption'. *Curr Top Behav Neurosci*. 2018;35:97-117.

Brimdyr K, Cadwell K, Widström AM, Svensson K, Phillips R. 'The effect of labor medications on normal newborn behavior in the first hour after birth: A prospective cohort study.' *Early Hum Dev*. 2019 May;132:30-36.

Buckley S, Uvnäs Moberg K. 'Nature and consequences of of oxytocin and other neurohormones during pregnancy and childbirth.' In: *Squaring the Circle: researching normal childbirth in a technological world*. Edited by Sheena Byrom and Soo Downe. Pinter & Martin, 2019.

Bystrova K, Ivanova V, Edhborg M, Matthiesen AS, Ransjö-Arvidson AB,

Mukhamedrakhimov R, Uvnäs Moberg K, Widström AM. 'Early contact versus separation: effects on mother-infant interaction one year later.' *Birth.* 2009 Jun;36(2):97-109.

Bystrova K, Matthiesen AS, Vorontsov I, Widström AM, Ransjö-Arvidson AB, Uvnäs Moberg K. 'Maternal axillar and breast temperature after giving birth: effects of delivery ward practices and relation to infant temperature.' *Birth.* 2007 Dec;34(4):291-300.

Bystrova K, Matthiesen AS, Widström AM, Ransjö-Arvidson AB, Welles-Nyström B, Vorontsov I, Uvnäs Moberg K. 'The effect of Russian Maternity Home routines on breastfeeding and neonatal weight loss with special reference to swaddling.' *Early Hum Dev.* 2007 Jan;83(1):29-39.

Bystrova K, Widström AM, Matthiesen AS, Ransjö-Arvidson AB, Welles-Nyström B, Wassberg C, Vorontsov I, Uvnäs Moberg K. 'Skin-to-skin contact may reduce negative consequences of "the stress of being born": a study on temperature in newborn infants, subjected to different ward routines in St Petersburg.' *Acta Paediatr.* 2003;92(3):320-6.

Bystrova K, Widström AM, Matthiesen AS, Ransjö-Arvidson AB, Welles-Nyström B, Vorontsov I, Uvnäs Moberg K. 'Early lactation performance in primiparous and multiparous women in relation to different maternity home practices. A randomised trial in St Petersburg.' *Int Breastfeed J.* 2007 May 8;2:9.

Champagne FA, Meaney MJ. 'Transgenerational effects of social environment on variations in maternal care and behavioral response to novelty.' *Behav Neurosci.* 2007 Dec;121(6):1353-63.

Chi Luong K, Long Nguyen T, Huynh Thi DH, Carrara HP, Bergman NJ. 'Newly born low birthweight infants stabilise better in skin-to-skin contact than when separated from their mothers: a randomised controlled trial.' *Acta Paediatr.* 2016 Apr;105(4):381-90.

Colson S. *Biological Nurturing.* Pinter & Martin, 2019.

Cong X, Ludington-Hoe SM, Hussain N, Cusson RM, Walsh S, Vazquez V, Briere CE, Vittner D. 'Parental oxytocin responses during skin-to-skin contact in pre-term infants.' *Early Hum Dev.* 2015 Jul;91(7):401-6.

Christensson K, Nilsson BA, Stock S, Matthiesen AS, Uvnäs Moberg K. 'Effect of nipple stimulation on uterine activity and on plasma levels of oxytocin in full term, healthy, pregnant women.' *Acta Obstet Gynecol Scand.* 1989;68(3):205-10.

Christensson K, Cabrera T, Christensson E, Uvnäs Moberg K, Winberg J. 'Separation distress call in the human neonate in the absence of maternal body contact.' *Acta Paediatr.* 1995 May;84(5):468-73.

Dobolyi A, Cservenák M, Young LJ. 'Thalamic integration of social stimuli regulating parental behavior and the oxytocin system.' *Front Neuroendo-*

*crinol.* 2018 Oct;51:102-115.

Davidovic M, Starck G, Olausson H. 'Processing of affective and emotionally neutral tactile stimuli in the insular cortex.' *Dev Cogn Neurosci.* 2019 Feb;35:94-103.

Ekström A.C., Thorstensson S. 'Nurses' and midwives' professional support increases with improved attitude – design and effects of a longitudinal randomized controlled process-oriented intervention.' *BMC Pregnancy and Childbirth.* 2015 15: 275

Feldman R, Weller A, Sirota L, Eidelman AI. 'Testing a family intervention hypothesis: the contribution of mother-infant skin-to-skin contact (kangaroo care) to family interaction, proximity, and touch.' *J Fam Psychol.* 2003 Mar;17(1):94-107.

Feldman R, Eidelman AI. 'Skin-to-skin contact (Kangaroo Care) accelerates autonomic and neurobehavioural maturation in preterm infants.' *Dev Med Child Neurol.* 2003 Apr;45(4):274-81.

Fuchs AR, Romero R, Keefe D, Parra M, Oyarzun E, Behnke E. 'Oxytocin secretion and human parturition: pulse frequency and duration increase during spontaneous labour in women.' *Am J Obstet Gynecol.* 1991;165:1515-23.

Handlin L, Jonas W, Petersson M, Ejdebäck M, Ransjö-Arvidson AB, Nissen E, Uvnäs Moberg K. 'Effects of sucking and skin-to-skin contact on maternal ACTH and cortisol levels during the second day postpartum-influence of epidural analgesia and oxytocin in the perinatal period.' *Breastfeed Med.* 2009 Dec;4(4):207-20.

Handlin L, Jonas W, Ransjö-Arvidson AB, Petersson M, Uvnäs Moberg K, Nissen E. 'Influence of common birth interventions on maternal blood pressure patterns during breastfeeding 2 days after birth.' *Breastfeed Med.* 2012 Apr;7(2):93-9

Hofer MA. 'Early relationships as regulators of infants' physiology and behaviour.' *Acta Paediatr Suppl.* 1994;397:9-18.

Jeannette C, Klaus PH, Klaus MH. 'No separation of mother and baby with unlimited opportunity for breastfeeding.' *J Perinat Educ.* 2004 Spring;13(2):35-41.

Jonas W, Wiklund I, Nissen E, Ransjö-Arvidson AB, Uvnäs Moberg K. 'Newborn skin temperature two days postpartum during breastfeeding related to different labour ward practices.' *Early Hum Dev.* 2007 Jan;83(1):55-62.

Jonas W, Nissen E, Ransjö-Arvidson AB, Wiklund I, Henriksson P, Uvnäs Moberg K. 'Short- and long-term decrease of blood pressure in women during breastfeeding.' *Breastfeed Med.* 2008 Jun;3(2):103-9.

Jonas W, Nissen E, Ransjö-Arvidson AB, Matthiesen AS, Uvnäs Moberg K. 'Influence of oxytocin or epidural analgesia on personality profile in breastfeeding women: a comparative study.' *Arch Womens Ment Health.* 2008 Dec;11(5-6):335-45.

Jonas K, Johansson LM, Nissen E, Ejdebäck M, Ransjö-Arvidson AB, Uvnäs Moberg K. 'Effects of intrapartum oxytocin administration and epidural analgesia on the concentration of plasma oxytocin and prolactin, in response to suckling during the second day postpartum.' *Breastfeed Med.* 2009 Jun;4(2):71-82.

Julius J, Beetz A, Kotrschal K, Turner D, Uvnäs Moberg K. 'Attachment to pets. An integrative view of human animal relationships with implications for therapeutic practice'. Hogrefe, Göttingen, Germany.

Kendall-Tackett K. 'Violence Against Women and the Perinatal Period: The Impact of Lifetime Violence and Abuse on Pregnancy, Postpartum, and Breastfeeding.' *Trauma, Violence, & Abuse,* 2007; 8:344-353.

Kendall-Tackett, KA Uvnäs Moberg K. 'Does synthetic oxytocin lower mothers' risk of depression and anxiety? A review of a recent study by Kroll-Desrosiers et al.' *Science & Sensibility* 2017

Kendall-Tackett KA, Uvnäs Moberg K. 'Pitocin, epidurals, and mothers' mental health. How can doulas help when mothers have high-intervention births?' *International Doula.* 2017; 25(3): 25

Keverne EB & Kendrick KM. 'Maternal behaviour in sheep and its neuroendocrine regulation.' *Acta Paediatr Suppl.* 1994; 397 :47-56.

Klaus MH, Jerauld R, Kreger NC, McAlpine W, Steffa M, Kennel JH. 'Maternal attachment. Importance of the first post-partum days.' *N Engl J Med.* 1972 Mar 2;286(9):460-3.

Lagercrantz H. The stress of being born.

Matthiesen AS, Ransjö-Arvidson AB, Nissen E, Uvnäs Moberg K. 'Post-partum maternal oxytocin release by newborns: effects of infant hand massage and sucking.' *Birth.* 2001 Mar;28(1):13-9.

McGlone F, Reilly D. 'The cutaneous sensory system.' *Neurosci Biobehav Rev.* 2010 Feb;34(2):148-59.

Neumann ID, Landgraf R. 'Tracking oxytocin functions in the rodent brain during the last 30 years: From push-pull perfusion to chemogenetic silencing.' *J Neuroendocrinol.* 2019 Mar;31(3):e12695.

Nielsen EI, Al-Saqi SH, Jonasson AF, Uvnäs Moberg K. 'Population Pharmacokinetic Analysis of Vaginally and Intravenously Administered Oxytocin in Postmenopausal Women.' *J Clin Pharmacol.* 2017 Dec;57(12):1573-1581.

Nissen E, Gustavsson P, Widström AM, Uvnäs Moberg K. 'Oxytocin, prolactin, milk production and their relationship with personality traits in women after vaginal delivery or Cesarean section.' *J Psychosom Obstet Gynaecol.* 1998 Mar;19(1):49-58.

Nissen E, Widström AM, Lilja G, Matthiesen AS, Uvnäs Moberg K, Jacobsson G, Boréus LO. 'Effects of routinely given pethidine during labour on infants' developing breastfeeding behaviour. Effects of dose-delivery time interval and various concentrations of pethidine/norpethidine in cord plasma.' *Acta Paediatr.* 1997 Feb;86(2):201-8.

Nissen E, Uvnäs Moberg K, Svensson K, Stock S, Widström AM, Winberg J. 'Different patterns of oxytocin, prolactin but not cortisol release during breastfeeding in women delivered by caesarean section or by the vaginal route.' *Early Hum Dev.* 1996 Jul 5;45(1-2):103-18.

Nissen E, Lilja G, Widström AM, Uvnäs Moberg K. 'Elevation of oxytocin levels early post partum in women.' *Acta Obstet Gynecol Scand.* 1995 Aug;74(7):530-3.

Nissen E, Lilja G, Matthiesen AS, Ransjö-Arvidsson AB, Uvnäs Moberg K, Widström AM. 'Effects of maternal pethidine on infants' developing breast feeding behaviour.' *Acta Paediatr.* 1995 Feb;84(2):140-5.

Odent M. *Childbirth and the Evolution of Homo Sapiens.* Pinter & Martin, 2013

Odent M. *Do We Need Midwives?* Pinter & Martin, 2015

Olausson H, Wessberg J, Morrison I, McGlone F, Vallbo A. 'The neurophysiology of unmyelinated tactile afferents.' *Neurosci Biobehav Rev.* 2010 Feb;34(2):185-91.

Olza Fernandez I, Uvnäs Moberg K, Ekström-Bergström A, Villarmea S, Hadjigeorgiou E, Leahy-Warren P, Nieuwenhuijze M, Kazmierczak M, Spidiruou A, Karlsdottir I, Buckley S. 'Birth as a neuro-psycho-social event: an integrative understanding of maternal experiences and their relation to neurohormonal aspects during childbirth.' *PLoS One.* Submitted.

Petersson M, Uvnäs Moberg K, Nilsson A, Gustafson LL, Hydbring-Sandberg E, Handlin L. 'Oxytocin and Cortisol Levels in Dog Owners and Their Dogs Are Associated with Behavioral Patterns: An Exploratory Study.' *Front Psychol.* 2017 Oct 13;8:1796

Pohl TT, Young LJ, Bosch OJ. 'Lost connections: Oxytocin and the neural, physiological, and behavioral consequences of disrupted relationships.' *Int J Psychophysiol.* 2019 Feb;136:54-63.

Putnam PT, Young LJ, Gothard KM. 'Bridging the gap between rodents and humans: The role of non-human primates in oxytocin research.' *Am J Primatol.* 2018 Oct;80(10):e22756.

Ransjö-Arvidson AB, Matthiesen AS, Lilja G, Nissen E, Widström AM, Uvnäs Moberg K. 'Maternal analgesia during labor disturbs newborn behavior: effects on breastfeeding, temperature, and crying.' *Birth.* 2001 Mar;28(1):5-12.

Sato A, Sato Y & Schmidt RF. 'The impact of sensory input on autonomic functions.' *Rev Physiol Biochem Pharcol.* 1997; 130:1-328.

Scott KD, Klaus PH, Klaus MH. 'The obstetrical and postpartum benefits of continuous support during childbirth.' *J Womens Health Gend Based Med.* 1999 Dec;8(10):1257-64. Review.

Sjögren B, Widström AM, Edman G, Uvnäs Moberg K. 'Changes in personality pattern during the first pregnancy and lactation.' *J Psychosom Obstet Gynaecol.* 2000 Mar;21(1):31-8.

Spengler FB, Schultz J, Scheele D, Essel M, Maier W, Heinrichs M, Hurlemann R. 'Kinetics and Dose Dependency of Intranasal Oxytocin Effects on Amygdala Reactivity.' *Biol Psychiatry.* 2017 Dec 15;82(12):885-894.

Stuebe AM, Michels KB, Willett WC, Manson JE, Rexrode K, Rich-Edwards JW. 'Duration of lactation and incidence of myocardial infarction in middle to late adulthood.' *Am J Obstet Gynecol.* 2009 Feb;200(2):138.e1-8.

Strathearn, L, Fonagu P, Amico J & Montague PR. 'Adult attachment predicts maternal brain and oxytocin response to infant cues.' *Neuropsychopharmacology.* 2009;34(13):2655-66.

Takahashi Y, Jonas W, Ransjö-Arvidson AB, Lidfors L, Uvnäs Moberg K, Nissen E. 'Weight loss and low age are associated with intensity of rooting behaviours in newborn infants.' *Acta Paediatr.* 2015 Oct;104(10):1018-23

Uvnäs Moberg K, Widström AM, Marchini G, Winberg J. 'Release of GI hormones in mother and infant by sensory stimulation.' *Acta Paediatr Scand.* 1987 Nov;76(6):851-60. Review.

Uvnäs Moberg K. 'The gastrointestinal tract in growth and reproduction.' *Sci Am.* 1989 Jul;261(1):78-83.

Uvnäs Moberg K, Widström AM, Werner S, Matthiesen AS, Winberg J. 'Oxytocin and prolactin levels in breast-feeding women. Correlation with milk yield and duration of breast-feeding.' *Acta Obstet Gynecol Scand.* 1990;69(4):301-6.

Uvnäs Moberg K. *The oxytocin factor. Tapping the hormone of calm, love and healing.* Pinter & Martin, 2011.

Uvnäs Moberg K. *The hormone of closeness. The role of oxytocin in close relationships.* Pinter & Martin, 2009, 2013.

Uvnäs Moberg K. *Oxytocin, the biological guide to motherhood.* Praeclarus Press, 2015.

Uvnäs Moberg K, Handlin L, Petersson M. 'Self-soothing behaviors with particular reference to oxytocin release induced by non-noxious sensory stimulation.' *Front Psychol.* 2015 Jan 12;5:1529.

Uvnäs Moberg K, Streit U, Nantke S. 'Oxytocin-Stoffwechsel. Körperkontakt unter erschwerten bedingungen.' *Deutsche Hebammen Zeitschrift.* 2017; 69(11).

Uvnäs Moberg K, Kendall-Tackett KA. 'The mystery of D-MER: What hormonal research can tell us about dysphoric milk-ejection reflex.' *Clinical Lactation.* 2018.

Uvnäs Moberg K, Buckley S. 'Oxytocin – a central hormone of labour and birth.' In: *Safety in Childbirth, Safety and Childbirth.* Edited by Sandra Moreno. The Italian Medical Association, 2019.

Uvnäs Moberg K, Ekström-Bergström A, Berg M, Buckley S Pajalic Z, Hadjigeorgiou E, Kotłowska A, Lenger L, Kielbratowska B, Leon-Larios F, Meier-Magistretti C, Downe S, Lindström., B., Dencker A. 'Maternal plasma levels of oxytocin during physiological childbirth – a systematic review with implications for uterine contractions and central actions of oxytocin.' *BMC Pregnancy and Childbirth.* 2019; In press.

Uvnäs Moberg K, Ekstrom-Bergstrom A, Buckley S, Massarotti C, Luegmair K, Kotlovska A, Grylka-Baeschlin, S Leahy-Warren P, Lengler L, Hadjigeorgiu E, Olza Fernandez I, Villarmea S, Pajalic Z. 'Oxytocin levels and oxytocin related effects in response to breastfeeding and consequences of medical interventions.' *PLoS One.* Submitted.

Uvnäs Moberg K, Handlin L, Kendall-Tackett K, Petersson M. 'Oxytocin is a principal hormone that exerts part of its effects by active fragments.' *Med Hypotheses.* 2019 Sep 6;133:109394. doi: 10.1016/j.mehy.2019.109394. [Epub ahead of print]

Uvnäs-Moberg K, Ekström-Bergström A, Berg M, Buckley S, Pajalic Z, Hadjigeorgiou E, Kotłowska A, Lengler L, Kielbratowska B, Leon-Larios F, Magistretti CM, Downe S, Lindström B, Dencker A. 'Maternal plasma levels of oxytocin during physiological childbirth - a systematic review with implications for uterine contractions and central actions of oxytocin.' *BMC Pregnancy Childbirth.* 2019 Aug 9;19(1):285. doi: 10.1186/s12884-019-2365-9.

Walker SC, Trotter PD, Swaney WT, Marshall A, Mcglone FP. 'C-tactile afferents: Cutaneous mediators of oxytocin release during affiliative tactile interactions?' *Neuropeptides.* 2017 Aug;64:27-38.

Walum H, Young LJ. 'The neural mechanisms and circuitry of the pair bond.' *Nat Rev Neurosci.* 2018 Nov;19(11):643-654.

Velandia M, Matthisen AS, Uvnäs Moberg K, Nissen E. 'Onset of vocal interaction between parents and newborns in skin-to-skin contact immediately after elective cesarean section.' *Birth.* 2010 Sep;37(3):192-201.

Velandia M, Uvnäs Moberg K, Nissen E. 'Sex differences in newborn interaction with mother or father during skin-to-skin contact after Caesarean section.' *Acta Paediatr.* 2012 Apr;101(4):360-7

Widström AM, Winberg J, Werner S, Hamberger B, Eneroth P, Uvnäs Moberg K. 'Suckling in lactating women stimulates the secretion of insulin and prolactin without concomitant effects on gastrin, growth hormone, calcitonin, vasopressin or catecholamines.' *Early Hum Dev.* 1984 Sep;10(1-2):115-22.

Widström AM, Ransjö-Arvidson AB, Christensson K, Matthiesen AS, Winberg J, Uvnäs Moberg K. 'Gastric suction in healthy newborn infants. Effects on circulation and developing feeding behaviour.' *Acta Paediatr Scand.* 1987 Jul;76(4):566-72.

Widström AM, Winberg J, Werner S, Svensson K, Posloncec B, Uvnäs Moberg K. 'Breast feeding-induced effects on plasma gastrin and somatostatin levels and their correlation with milk yield in lactating females.' *Early Hum Dev.* 1988 Mar;16(2-3):293-301.

Widström AM, Marchini G, Matthiesen AS, Werner S, Winberg J, Uvnäs Moberg K. 'Nonnutritive sucking in tube-fed preterm infants: effects on gastric motility and gastric contents of somatostatin.' *J Pediatr Gastroenterol Nutr.* 1988 Jul-Aug;7(4):517-23.

Widström AM, Matthiesen AS, Winberg J, Uvnäs Moberg K. 'Maternal somatostatin levels and their correlation with infant birth weight.' *Early Hum Dev.* 1989 Dec;20(3-4):165-74.

Widström AM, Wahlberg V, Matthiesen AS, Eneroth P, Uvnäs Moberg K, Werner S, Winberg J. 'Short-term effects of early suckling and touch of the nipple on maternal behaviour.' *Early Hum Dev.* 1990 Mar;21(3):153-63.

Widström AM, Christensson K, Ransjö-Arvidson AB, Matthiesen AS, Winberg J, Uvnäs Moberg K. 'Gastric aspirates of newborn infants: pH, volume and levels of gastrin- and somatostatin-like immunoreactivity' *Acta Paediatr Scand.* 1988 Jul;77(4):502-8.

Widström AM, Werner S, Matthiesen AS, Svensson K, Uvnäs Moberg K. 'Somatostatin levels in plasma in nonsmoking and smoking breast-feeding women.' *Acta Paediatr Scand.* 1991 Jan;80(1):13-21.

# Index

*Available from Pinter & Martin*
*in the* ***Why it Matters*** *series*

1. ***Why Your Baby's Sleep Matters*** Sarah Ockwell-Smith
2. ***Why Hypnobirthing Matters*** Katrina Berry
3. ***Why Doulas Matter*** Maddie McMahon
4. ***Why Postnatal Depression Matters*** Mia Scotland
5. ***Why Babywearing Matters*** Rosie Knowles
6. ***Why the Politics of Breastfeeding Matter*** Gabrielle Palmer
7. ***Why Breastfeeding Matters*** Charlotte Young
8. ***Why Starting Solids Matters*** Amy Brown
9. ***Why Human Rights in Childbirth Matter*** Rebecca Schiller
10. ***Why Mothers' Medication Matters*** Wendy Jones
11. ***Why Home Birth Matters*** Natalie Meddings
12. ***Why Caesarean Matters*** Clare Goggin
13. ***Why Mothering Matters*** Maddie McMahon
14. ***Why Induction Matters*** Rachel Reed
15. ***Why Birth Trauma Matters*** Emma Svanberg
16. ***Why Oxytocin Matters*** Kerstin Uvnäs Moberg

***Series editor:*** Susan Last

*pinterandmartin.com*